Richard Nichols is c[illegible] Britain's most popul[illegible] played a major part i[illegible] the public. In autumn [illegible] Britain's first CB mag[illegible] Richard Nichols has ac[illegible] legalisation of CB on na[illegible] radio, on television and in the press. He is a [illegible]mber of the CBA (Citizen's Band Association), the UKCBC (United Kingdom Citizen's Band Committee) and REACT (Radio Emergency Associated Citizen's Teams).

CB RADIO: A HANDBOOK

Richard Nichols

A STAR BOOK
published by
the Paperback Division of
W. H. ALLEN & Co. Ltd

A Star Book
Published in 1981
by the Paperback Division of
W. H. Allen & Co. Ltd
A Howard and Wyndham Company
44 Hill Street, London W1X 8LB

Fourth edition

Printed in Great Britain by
Anchor Press, Tiptree, Colchester, Essex

At the time of going to press CB radio is illegal in the UK and the Publishers would like to make it clear that it is not our intention to encourage or approve of any illegal activity.

Contents

Introduction

Radio communication is not all that new as a concept, and has been with us in various forms for many years. In the UK at least it is traditionally regarded as being the exclusive preserve of Government branches and various other forms of Authority in all its guises. The idea that each private individual is entitled to use of the airwaves for whatever law-abiding purpose he or she may see fit is, however, relatively new. Yet that is all that Citizen's Band Radio (or CB, as it is popularly known) was ever designed for.

The idea is that individuals are entitled to have in their car, home, boat or caravan a simple two-way radio set on which they may discuss the weather or other topics with similarly-equipped people. Such sets are necessarily made as cheap and easy to use as current technology allows.

This concept was readily accepted and adopted in America as long ago as 1948 and has since spread to more than 60 other nations. In Britain it produced a wave of near-hysteria in government departments, and strenuous attempts were made to deny the facility to the British public. In the main these efforts were centred around a straightforward refusal to acknowledge the benefits which such a service could offer, coupled with a number of official statements which purported to explain exactly why the thing was so undesirable. Because the proponents of CB for Britain were already (illegally) using imported American equipment part of the official denial was ready-made. CB, we were told, interferes with a number of existing services to an unacceptable level. Services suffering from CB pirates were often said to include hospital paging systems, model aircraft control and domestic TV sets.

The use of incautious wording in statements to the media gave rise to a fairly widespread belief that everyday life in Britain would be brought to a standstill immediately upon the introduction of a CB facility. It was even suggested that heart pacemakers would stop dead and the populace would turn into bank robbers, mass murderers and rapists overnight.

Subsequent examination of the situation has revealed that this is not actually the case, and indeed life may well be actually enriched by CB rather than despoiled. No need to lock up your daughters after all. In fact the Government has now said that it finds CB to be a Good Thing, and hopes to be able to find a suitable frequency for it quite soon. The fact that only two years ago there were 'no frequencies' which might be allocated to CB is a reasonably illuminating guide to the nature and validity of the rest of the official disclaimers.

But now that we know there's nothing wrong with it after all we can get on with the business of finding out about it and getting it working. You'll probably need some help, and here it is.

When dealing with matters of a specialised nature, such as electrical or mechanical devices, it is better not to know. In fact the less you know the better off you are. Exceptions to this rule must necessarily be made in the cases of people like Edison, Baird and Marconi, who by virtue of being born in close proximity to places or events of importance have it as their bounden duty to enquire into The Nature of Things, and invent the odd item from time to time. This is perfectly natural, and is known as Progress. It does not, however, free the rest of us from the responsibility of ignorance. It still does not require a complete understanding of Mr Baird's invention to be able to switch off *Crossroads*, which is only right and proper.

In this splendid frame of mind plans for writing a book containing everything you need to know about CB were formulated. After much inner struggling and agonising selection it became clear that there were only six facts essential to the operation. Once again informed sources interceded, this time in the shape of Mike Bailey at W. H. Allen, who intimated that such a book might break at least two publishing records. It would be the shortest book in history—and the author would be paid less than anyone before and very likely since.

Serious consideration of this intelligence has led to the regrettable inclusion of a vast and unpalatable amount of spurious information about CB, largely of a distressingly technical nature, for which I apologise. Strenuous efforts have been made to keep the extra bits as simple as possible, and although there is a fairly substantial amount of whying and whereforeing it is still not completely indigestible, and no attempt has been made to delve any deeper into the scheme of things than is absolutely necessary. The basic six facts still remain, of course, and students of simplistics will have little bother in spotting them and discarding everything else. Also, in an attempt to alleviate the general gloom, the text contains two jokes, both of which will escape all but the most dedicated simpleton, which is as it should be. What will be immediately noticed by anyone who isn't already familiar with CB is that there are a great number of words and phrases which are of a reasonably technical disposition and therefore totally incomprehensible, at least to begin with. There is a fairly substantial glossary of terms at the back of the book, and that should explain everything as you go along. On the other hand, of course, it may not, in which case one of us hasn't been paying attention. I just hope it's you . . .

Richard Nichols, February 1981.

1

A history of CB and the English pirates

It's a fact that the process of invention is necessarily a slow one, in which very often the inventor receives all the credit when some of it may rightfully be due elsewhere. Commonly the Wright brothers are regarded as being responsible for powered flight. But the knowledge which enabled them to make that first flight had existed for some time; the only thing missing was an engine small enough and light enough to get them off the ground. Since contemporary motor car engines were unsuitable a man named Charles Taylor built from scratch a petrol engine which was based on someone else's earlier design for a steam engine. Taylor took six weeks to build the engine which finally, in 1904 lifted the "Flyer" off the sands of Kittyhawk and into the history books, although not into the American ones until 1942, when the US finally recognised the brothers' achievement.

The Wright brothers are not the only people whose final results have been dependent upon the efforts of others. Many inventors have gone ahead with designs for things which were dependent on components which themselves had not yet been invented. Still more made their greatest discoveries by accident—X-rays and penicillin, for example. It was such a combination of chance circumstances which led to the sudden and almost worldwide CB boom of the early seventies and gave us all an electronic gadget which has generated more interest than anything since the advent of colour TV.

The story began shortly after the war when the American government legislated for the introduction of the General Mobile Radio Service (GMRS) on the VHF frequency of

467MHz, which was to be a high-grade two-way facility which would be freely available to the general public. This in itself was a far-seeing step as it was already moving beyond the traditional "ham" services which existed in many countries.

The major problem with GMRS was that although there was a will there was not yet a way; existing valve-radio technology produced large, cumbersome devices which consumed staggering amounts of electricity and were so fragile they could be broken by a careless sneeze.

The major breakthrough which came to the rescue was the transistor, which made its world debut in 1948. This miniature device did all that a valve could do on a smaller scale. Although it was by no means as delicate as the valve the early transistors could hardly be described as robust and it was nearly ten years before they began to appear in car radio sets. But by then the CB situation had changed considerably.

It is a fact that it becomes technically more complex to transmit a radio signal over a given distance as the frequency used is raised. In addition it requires more power. Thus a 5 watt signal at 10MHz will travel further than a 5 watt signal at 500MHz. Despite the arrival of the transistor, GMRS was ahead of its time and ahead of contemporary technology.

So it was that in 1958 the Americans introduced what they described as "Class D Citizen's Band Radio". This service was designed around the existing electronic technology and provided 23 channels in the 27MHz band for private citizens to talk to each other, either from their homes or their cars.

CB, as the service quickly became known, was initially ignored. To anyone who heard the assorted hoots and whistles of these 1958-built radio sets this will not come as a surprise. It was some time before technology caught up with ideas and together formed progress; almost certainly, as far as electronics are concerned, the bulk of the development came in the form of spin-offs generated by Mr Neil Armstrong's urgent desire to get away from it all. After running up the largest travel bill in history he finally made it to the moon in 1969 after a ten-year period of development which had ushered in solar cells, computers and pocket calculators. With one exception the first wave of the micro chip revolution had arrived.

This revolution finally enabled the CB facility to get off the ground in the manner to which it was shortly to become drastically unaccustomed. To begin with, however, all it meant was that people using the early sets threw them away and bought newer ones on which they could hear each other speak rather better than they could hear electronic Percy Edwards impressions. Even so there was never what you could call a CB boom. If anything it was a sort of damp puff, largely confined to truckers.

America, as students of geography will readily confirm, consists principally of vast tracts of nothing interspersed with small lumps of frenzied development which the inhabitants are pleased to call towns. The truckers' job is to thunder between these places carrying vast loads of comestibles and the fragmentia of civilised life. A trucker can quite easily go for hours without seeing another soul, never mind talking to anyone. There has always been a kind of mystic brotherhood associated with this way of life and it's no surprise to discover that the truckers stick very closely together. The advent of CB meant that they could do this much more effectively, even while they were travelling in opposite directions, and it's hard for anyone who hasn't experienced the loneliness of the open road to realise exactly what this must have meant to them.

CB also gave them much assistance of a highly practical nature; many of the truckers are self-employed owner-drivers and every second of delay costs money. With CB they need never waste time looking for their delivery location in a strange town or wait for hours on a highway for assistance after a breakdown. Even in towns and cities CB helped, as anyone who has missed a turning and tried to turn round in a forty foot artic unit will know.

Despite all that CB still hadn't taken off in a big way; it was a cult even then, but very much a minority one. What finally gave it the lift it needed was the fuel crises of the early seventies. All Americans, until then, had taken it as their God-given right that the world was designed for their benefit. This misapprehension applied particularly to oil, and a plentiful supply of cheap fuel had always been part of the American heritage, never mind the American dream. American motor manufacturers had naturally built cars on exactly the same premise, and their gas-guzzling monsters surely qualified them for a special "friend of OPEC" award all their own. They got it in 1973, when the fuel pumps suddenly ran dry and the cars began to grind to a halt.

Apart from near-panic, this induced two major changes to the American motorist and his way of life. First, and most obvious, was the scarcity of fuel; drivers had to go looking for it instead of filling up at the station which washed their windows best. Second, and far more sinister, was the introduction of a blanket 55mph speed limit—the hated double nickel.

Truckers were obviously hardest hit by all this—they needed fuel to survive, but were frequently compelled to make journeys down long, unfamiliar roads and into the wide open spaces, and they also made more money if they got from place to place as quickly as they could. It was therefore a natural progression for them to use their radio sets to keep going; firstly to find fuel and secondly to warn each other of police speed traps—smokey reports, as they are now known. At this moment the CB became a necessity for the truckers rather than the luxury it had been. At the same time the rest of the American motorists twigged that CB was going to be quite useful. Americans never do anything by halves, and once they'd decided that CB was going to be a good thing they piled in at a tremendous rate and the boom really got going.

In January of 1977 the original 23 channel allocation had to be increased to 40 and at the same time the specifications for CB rigs were advanced slightly. The rules were that rigs made to the old spec had to be off the dealer's shelves by January 1, but the information was given well in advance, so the closing months of 1976 saw the biggest nationwide clearance sale in history. The old sets were so cheap that everyone could afford one, so everyone went out and bought one, if not several.

By this time, of course, the Japanese had established their reputation for low-priced electronic hardware, also their inscrutable ability to churn the stuff out at an amazing rate. They did just that with CB rigs, meeting the demand for ten million mobile units in 1977 largely by themselves. In the way of all new fads and crazes the demand was at its greatest at the beginning and then it settled down. Unlike most fads and crazes it did not die out completely, but finally pegged at a steady two million units a year. Most of these will probably be replacement sets rather than first-time purchases, especially now as the third-generation sets start to arrive in the shops. Even so there may be as many as forty million Americans on the air, perhps half of whom regard the airwaves with the same proprietorial feeling as they do oil, and consequently haven't paid their licence fee.

Naturally enough, with all these people looking for a bit of the 27MHz band to talk on, there have to be rules and although most of them are dictated by common sense they are set out in the licence by the FCC (Federal Communications Commission) for all to digest. The truckers have been using their CB for so long that by the time middle America got on the air there was already a patois of the airwaves which required a dictionary to follow it, at least for the beginner. Trucker talk mixed on the airways with the official ten-code, as used by the emergency services, the CB ten-code, the ham radio Q-code and a whole lot of other codes which were unofficial to say the least. Despite all this the early exponents of CB may be forgiven for not realising that they were participating in the birth of a legend; as far as they were concerned they had found a new gadget which promised to be of so much practical use that they quickly discovered that once started, they could not live without it. Thus it was that CB developed this dual role—on the one hand it had so much practical value, on the other it appealed to a sense of fun and adventure.

To the motorist CB was the answer to many prayers. Used

widely, it provided advance warning of traffic holdups, bad weather and all manner of motoring information. It didn't take American officialdom long to spot all this, and a survey in Ohio revealed that CB was saving life perhaps 500 times every year. Firstly, by warning of holdups on the freeway it was avoiding the multiple pileups which we call motorway madness; this kind of saving is hard to quantify. Secondly, drivers were using their rigs to summon emergency services to the scene of road accidents. Anyone concerned with the giving of immediate first aid to injured people will tell you that in cases of severe injury the help given in the first few moments can mean the difference between life and death to the victim. Some years ago the ambulance service around Nottingham claimed that they could respond to a call in any part of their area within fourteen minutes at the outside. That's after they've had the call. Fourteen minutes is three times longer than it takes to bleed to death from serious injury; what happens if it's already taken ten minutes to find a phone that works?

With CB the call is more or less instantaneous, and in the fourteen-minute waiting period a complete ignoramus can be given detailed instruction on how to stop bleeding and so on by an expert who need be no closer than twenty miles. It is easy to see how life can be saved. In 1979, the last year for which figures are available, 6,352 people died in road accidents in this country. We'll never know how many of them might have lived if CB had been available.

For some reason CB had never been made legal in the UK. Many countries watched the American experience with interest and as a result introduced CB facilities of their own on similar lines, but Great Britain wasn't one of them. By late 1980 there were more than sixty countries with legal CB, some of them behind the iron curtain. Nearly every country in Europe had got it—France at the very end of 1980 and then unwillingly, in response to widespread illegal use throughout the country, much in the same way that the Australian Government were forced to legalise in 1979.

In 1948, when the transistor was being invented and the

Americans were putting it to good use, we in Britain were taking steps to ensure that it would never be used over here if it was possible to prevent it. The instrument of obstruction was the Wireless Telegraphy Act which came into force in 1949. This splendid piece of legislation was primarily aimed at ensuring that the state-owned GPO maintained a total monopoly in the field of communication. The act made it a criminal offence to use a radio receiver or transmitter without a licence. Effectively this meant that you had to give the government money before you could listen to the Palm Court Orchestra or watch the very early TV shows. However it also meant that you couldn't use a transmitter without paying the government money as well. If you wanted to talk to someone on a radio you went to the Post Office, paid your licence fee and off you went . . . well, no. Actually it wasn't that simple. Although the GPO were put in charge of the administrative side of things, they could only issue licences which they were authorised to by the Home Office. The prerogative to dispense the radio spectrum belongs to Her Majesty's Minister for Home Affairs—the Home Secretary. Once he had released part of the spectrum for a given use the GPO would then issue licences. The only type available to the private individual was the "Ham" licence, which required the passing of some fairly stringent exams as well as a working knowledge of Morse Code. Exactly why you might need Morse Code during a conversation on radio any more than you would on the telephone is not clear, but presumably it was not done simply to keep the number of Ham licences as low as possible.

No government, of any denomination, saw fit to make any change to this situation for some considerable time. When change was made it was nearly twenty years later in 1968. This year saw several advances in radio administration, beginning with the Post Office Incorporation Act (during which they lost their General) and the now-famous Statutory Instrument No. 61, 1968, Radiotelephonic Transceivers (Control of Importation and Manufacture). This staggering document made it specifically illegal to manufacture or import radio equipment operating in the 27MHz band. This surprising and drastic step went largely unnoticed in this country, mostly because no one over here had got the faintest idea what 27MHz radio equipment was. It would have been quite understandable if this sort of legislation had been introduced in the late seventies, when the CB boom was just beginning over here, but in 1968 it was introduced in response to nothing, or so it seems. The most logical explanation for it lies with the domestic electronics industry who could probably see far enough ahead to realise that CB represented a threat to their existing market and indulged in some fairly hefty lobbying.

It's unlikely that anyone in Britain at the time saw the way that CB was going to boom—after all, it hadn't yet done so in the States—but many of the electronics firms were in the very lucrative market for Private Mobile Radio users. The PMR licence entitled business concerns to licence a number of sets on an allocated frequency for their own use. Because the number of sets used by any one concern was usually not very large the price of each unit was remarkably high, and traditionally British industry has enjoyed inflated prices for radio sets of this nature. CB sets operating on 27MHz and made in America or Japan were priced ridiculously cheaply by comparison. Anyone with half a mind could have brought in rigs from abroad and saved himself a fortune by buying foreign, and it seems likely that the Statutory Instrument was designed to prevent this at the commercial level rather than prevent Britain taking to the airwaves.

But although no one in Britain really knew what CB was the first glimmerings were beginning to seep through and various odd bods were beginning to talk to each other on the air. Probably the first CB sets which reached this country were the 100-milliwatt-output hand-held variety which require no licence in the States. Until the Instrument of 1968 these were appearing from time to time in the shops as children's toys, but that soon stopped. The first recorded British CB activity was on these toy-like radio sets. A 100-milliwatt hand-held has a range of no more than a few hundred yards on a good day with a following wind, but as early as 1965 a group of afficionados known as the Charlie Bravo group were swanning about London chatting to each other for some time. Originally they worked channel 11, but interference forced them up to channel 14 where they stayed for some time. The low power of these sets probably accounted for the somewhat charmed life of the group, but eventually a number of prosecutions forced them off the air in the early seventies. One of these early casualties was actually transmitting on a hand-held as he walked past Waterloo House, home of the Radio Regulatory Department of the Home Office; someone ran out and nabbed him. Talk about naive.

The Charlie Bravos were soon replaced, however, by the Lima Echo group. This lot were considerably more sophisticated all round, and used the 23-channel mobile rigs which were then all that was available—the 40-channel allocation was still some years off, even in the States. The Lima Echo group took up where Charlie Bravo had left off and adopted channel 14 as their net channel. At the time there weren't enough of them for it to be regarded as a calling channel, but as the ranks swelled they stayed on 14, despite the fact that the Americans have consistently used 19. Lima Echo never did get driven off the air—their ranks simply grew and grew until there were so many breakers on channel it was impossible to say who was who.

It's hard to see how anything like CB could have grown to the extent it has, when it was illegal and without help from either the media or from an industry anxious to push its wares. But, indisputably, it grew. Perhaps the thing which contributed more to the growth of British CB than anything else came from a Country and Western singer by the name of C W McCall. His only real claim to fame came in the form of a single released just after the fuel shortages had caused CB to take off in the States. *Convoy* had little to recommend it on a musical basis and not much in the way of lyrics. Nevertheless it was a monster hit, relying heavily on the mysterious and intriguing jargon of the American truckers for its potency. If the PR executives of all the major CB manufacturers had clubbed together and put up a million pounds to be spent pushing CB to the masses they wouldn't have achieved one tenth of what good old CW did for them by accident.

The British press were absolutely fascinated by the jargon and acres of valuable newspaper columns were devoted entirely to explaining exactly what a smokey bear was, who Cabover Pete was and just why he had a reefer on. And if that wasn't enough, pregnant roller skates* completely demolished all media resistance to CW's otherwise mediocre little ditty. *Convoy* was such a blockbuster that several years after its release a film company felt obliged to commit it to celuloid—first the film of the book, then the book of the film, now the film of the record. And still no one knows what it was all about.

It was natural that once the record had been accepted as a cult item in its own right all aspiring long-distance truckers would seek to emulate their heroes. If the record was a good thing then it followed that CB was a good thing also.

Dear old Freddie Laker must also stand up and take his share of the blame at this point; perhaps if he hadn't made it so easy and so cheap to get to America from our sheltered isle there wouldn't have been quite so many inbound suitcases which weighed considerably more than they had on the outward journey. While airport Customs officers searched diligently for that extra 200 cigarettes CB rigs were flooding into the country. Perhaps, with hindsight, flooding is the wrong word; trickling would probably be better, although at the time it seemed like a flood.

By 1979 the Citizen's Band Association (CBA) was already in full swing, having been formed in 1976, and they were pressing the Home Secretary to introduce CB into the UK on the VHF frequency of 232MHz. The Home Office countered this by pointing out that 232 was already allocated. This was true, as the CBA were quick to agree, but it was allocated to the Lancaster Bomber, which had gone out of service in

* See Chapter 9—*Trucker-talk* for an explanation of these mystifying terms.

1958. Demand for "Lancaster Band" CB grew and grew, until finally the Home Office was forced to own up to the truth. Their attitude changed, and from saying that there were "no frequencies available which may be allocated to a service of this type" (a palpable lie, and astoundingly patronising into the bargain), the Home Office came out of its closet and admitted that they felt that "the disadvantages of such a service outweighed the advantages". By this time everyone with half a brain could see that there were definite advantages to CB, but no one except HMG could spot any of the disadvantages—nothing important, that is—and they weren't telling.

The problem here is one of communication. The Home Secretary is the only person who may speak with any authority on the subject of radio spectrum allocation and he is not generally regarded as the most accessible person in the whole world. Someone once said that the law, like the Ritz hotel, is open to everyone, and a similar principle may be applied to members of the Government. You may write to them if you wish. They may write back. Then again . . . About the only effective way of getting to grips with a minister is via the parliamentary question. Most of these get a written answer anyway, so there is still no certainty of being able to discuss a given subject with the minister concerned—even at the very top.

During the late seventies a great number of questions were put to HMG in the House and they all got written answers of varying length and stupidity. At one time the Government said that they were afraid that if a CB facility were introduced its primary use would be for "non-serious purposes". Clearly nobody in government had watched *Crossroads* or listened to Radio One, or they would have realised that vast amounts of airspace were already being used for purposes of a stupendously non-serious nature. Obviously HMG regarded the proponents of CB for this country as a few cranks who would easily be dissuaded by a simple process of stonewalling—something which all politicians are especially good at. At the time they may have been right. The earliest recorded writing on the subject of the CB pirates came in October 1978, when a London-based events magazine estimated that there were as many as 500 pirates operating in the capital. They were referring to the not-so-embryonic Lima Echo group. Less than a month later the estimate was revised upwards to 800. Many of these pirates congregated late at night to work some choice skip from the US, as the late autumn sunspot activity produced some really freaky DX contacts. The location most of these pirates chose was Alexandra Palace. Long ago the Beeb had figured out that this high ground was a prime site for a radio transmitter, and it was no accident that the first TV transmissions came from there. However the pirates weren't the second people to come to this conclusion, and the mast at Ally Pally concealed many other transmitters, several of them of a somewhat furtive nature. For example it was well-accepted that Special Branch had a station on the hill. It was also believed that this worked on a frequency not unadjacent to 54MHz—the second harmonic of 27MHz—which made it remarkably susceptible to interference from the pirates.

Despite this the police activity at Ally Pally was of an exceedingly gentlemanly nature, usually confined to obtaining polite assurances that while everybody was busy monitoring skip like fury no one was so crass as to actually transmit at all.

Meanwhile, back at the factory (if that's what MPs call the House in their private moments) things were hotting up. The government of the day were using Lord Wells-Pestell as their mouthpiece on matters pertaining to CB, and in this capacity he committed to Hansard and posterity some of the most splendid pieces of rubbish ever to have been heard in the long history of the noble institution in which he was privileged to speak. Replying to a question on the subject of CB, my Lord stated the case against with a devastating clarity which left his admiring audience gasping. The Government, he said, did not like the idea of having large armies of people within easy communication of each other. Clearly His Lordship was in for the shock of his life when he heard a telephone bell ringing. Equally we must assume that he was later left speechless after discovering the existence of the letter-post, since it is obvious that both of these state-controlled methods of easy communication were currently beyond the scope of his knowledge.

But His Lordship's specialised information on the subject of CB didn't stop there. He went on to intimate at still more horrific catastrophes in store for the unwary when he pointed out that the dim governmental view of these fellow-communicants was worsened by the fact that seemingly they all felt that they had "the right to rape, pillage and steal". Lord Wells-Pestell did not vouchsafe to the House how this attitude would be brought about in the breasts of innocent members of the public once they had taken two-way radio to their bosom (and presumably their cars as well). Neither did he attempt to explain why it should be that police officers, firemen, gas board officials and taxi drivers, all of whom used two-way radio in the course of their jobs, had not already embarked upon a trail of carnality and destruction. The AA is, naturally, a separate case.

Undeterred (or possibly spurred on) by the terrible consequences in store for them, the pirates kept on keeping on. Their numbers swelled. And as the number of illegal broadcasters grew, and their influence spread all over the country, instead of being confined to London, so the police activity stepped up. Right from the beginning it was believed that Hornsey police (very close to Ally Pally) had a rig in the station and their detailed knowledge of the CB scene indicated that they had been monitoring for some time, but with the growth in illegal use the police employed stronger tactics. To the early pirates the names of Goldfinger and Electric Lady are both well-known, and it was widely believed that they were police officers. Certainly they acted as bait for the unwary, luring them to eyeballs notable principally for the fact that they were attended by a lot of gentleman in pointed hats.

Examination of the situation with regard to CB revealed that there are only three offences with which transgressors may be charged: importation, connection to a power source and transmitting. Of these three, the last two are hard to prove unless the accused is willing to admit to them. The offence of transmitting may be proved if someone is caught with the microphone at his lips and his finger on the button, but that's about as likely as catching a bank robber with his hand in the till. Connecting to a power source should be easier, but here the offence is that of making the connection, not being in possession of a set which is so connected. Unless the pirate wishes to admit to the offence, or can be tricked into doing so, prosecution is not really viable. Which leaves illegal importation. This is the easiest one of the lot, because conviction depends only upon physical evidence of the set itself. This fact gave rise to some early confusion and was also the reason that the principals in the night-time hunts and arrests were officers of HM Customs and Excise. Customs officers enjoy wide-ranging powers, particularly of entry and search, and the goods in question were clearly subject to seizure by Customs because duty had never been paid on them. The power under which these sets were seized came from the Customs and Excise Acts of varying dates, and for a long time it was believed that because of this police officers were not empowered to seize CB rigs without the permission of the owner. So much confusion over this fact existed that eventually many police forces gave up chasing possible CB offenders—in fact it was not even clear whether the Road Traffic Act empowered officers to stop motorists on the road in order to check for possible illegal radio installations.

In the event it was obvious that possession of a 27MHz CB radio definitely constituted at least one offence, possibly three, and therefore a police officer would be entitled to seize it as evidence under common law, pending prosecution. But so much was happening so fast that much of the legal tangle became a pointless bickering.

But the persecution and prosecution of so many CB offenders only added to the myth that was making it so attractive to those on the outside. It was also having another, more noticeable effect. Once the pirates had accepted the fact of police and Post Office monitoring of the 27MHz band it was clear that a means of speaking which only the few could understand would be necessary.

Many of the pirates had already affected the trucker-talk of the USA for fun. Now it became a necessity. Even more necessary was the need to refer to people and places by some form of code name. The practise of adopting some form of handle had long been established and was now less humorous than it used to be. But place-names had to be altered as well. In the beginning the international phonetic alphabet sufficed, especially with regard to pub names (almost uniquely, all British motorists navigate by means of pubs rather than any other form of landmark) and one of the most famous watering holes for pirates was Delta Hotel. It didn't take the plods very long to figure out that it meant the Dutch House in South-East London, and it didn't take them long to get round there with a substantial group of bodies and make a substantial number of arrests.

It was about then that the pirates discovered that the Customs officers were entitled, in pursuance of their duties, to rend your vehicle into its component parts, with axes and crowbars if necessary, without being obliged to put it back together or compensate for any damage. Feelings began to run a little high for a while, but during 1980 it began to calm down again. There were several reasons for that.

One was the, by now, incredibly high number of pirates operating on 27MHz. From 500 in October '78 the figure had climbed as high as an estimated quarter of a million. It was clearly impossible to control that number of illegal users on the air, and the resources of the various long arms of Authority were stretched to the limit. Up until early 1980 the bulk of the police work had actually been done by Customs officers, but cuts in government spending meant that their resources and manpower were being stretched to impossible limits in this and other fields. The advent of Dutch Elm disease now meant that even trees had to be correctly documented and the drug problem was getting much worse at an alarming rate. Obviously the Customs people could not justify the allocation of considerable amounts of their dwindling manpower to chasing criminals who were doing nothing worse than talk to each other.

Further developments seemed to indicate a softening of attitude on the part of authority. The Home Office seemed at last to realise the ridiculous position they were in, decrying as impractical and useless something which had already been proved to be exactly the opposite in nearly every other country in the civilised world. They then went so far as to suggest that while CB might after all be a good thing there seemed to be nowhere to put it—the old chestnut about a lack of frequencies again. What they really meant, of course, was that they were looking for a practical alternative to 27MHz which would give the British electronics industry a head start over the nasty Japanese, who had warehouses full of 27MHz rigs poised for despatch to this country the minute the Government gave the word. In a market estimated to be about the same size as the hi-fi market in this country it didn't seem too unreasonable to reserve as much as possible of it for our own chaps. After all, it was for this very reason that our own chaps had put their weight behind the Statutory Instrument which kept the stuff out of the country. Unfortunately it also stopped them from making it, which meant the pirates were all using imported gear.

Meanwhile, as the search for the right frequency went on the search for loopholes also went on, and there were further rumblings of discontent as many individuals searched through the reams of paper which constituted the agreements with the EEC in the fervent hope that something contained within could force the hand of HMG. For a time their Treaty of Rome seemed to offer the most hope, since it forbade member countries to discriminate against imports from fellow-members except on the grounds of national policy.

But the real change came about with the change of government. The Conservative Party, who had, while in Opposition, said that they felt CB to be of some benefit, came to power in May of 1979, and it was after that event that the official attitude began to soften. In late 1979 the GLC produced their own discussion document on CB, rather as if they were going to introduce CB into London without further reference to the Government. Response to their general enquiry was limited to a high percentage of voices raised in favour and very few against. The GLC discussion document preceded the Government Green Paper by only two months, and that was its sole claim to competition, for it made no suggestions nor did it offer any support—other than to say that if the people of London were in favour of CB then they would make representations to the Government accordingly. But by then the Green Paper was expected anyway, so it was all fairly irrelevant.

The Green Paper finally came out into the open in August of 1980 and had long been expected to put an end to the piracy that had snowballed over the past couple of years. Many hoped that the Green Paper would announce the swift adoption of the American 27MHz system which was well-proven and for which the technology already existed, even if the frequency wasn't ideal. The CBA still hoped that their plans for a VHF frequency would be adopted, even at this late stage, but in the final event the Government produced proposals for something which they chose to call "Open Channel" and which worked on a frequency of 928MHz. This unfortunately provided for a service which offered as much range as two cocoa tins and a piece of string at a price approximately three times greater than the existing 27MHz sets. This was immediately greeted with almost universal scorn, and although the Government had allowed until the end of November for the public to make their opinions on the subject known, the papers the day after the Open Channel proposals were announced really said all there was to say on the subject.

The hope that the Green Paper would put an end to piracy and controversy was dashed almost straight away and the large numbers of pro-CB people who had been holding back from piracy in the hope that they could get the legal feeling in 1981 seemed to rush out and start a mini CB boom all of their own towards the end of 1980. Estimates of the numbers of pirates have always only been estimates, but they seem to have doubled with great rapidity; it seemed that they could not go on doubling forever. But in late 1980, as the French Government collapsed under the pressure of the illegal users and scrapped their plans for 928 in favour of 27MHz, the number of pirates in the UK had gone from the midsummer guess of 250,000 to a staggering 500,000. The great British idea of having our own CB frequency and keeping the foreign competition out had failed—while the Government had its back turned the Japanese had enjoyed most of the pickings of the British CB market, and instead of keeping them out and boosting our domestic industry we did the exact opposite.

2

Burners, sliders and Outbanders—the American problem

It would be easy for anyone in Britain to think that we are the only country which has had the misfortune to suffer the slings and arrows of a shortsighted policy on allocation of the radio spectrum, but this is not the case. Indeed the British pirates were following a fairly well-defined path which other countries had already experienced. Everyone seems to be aware that Australia was the first country to suffer from an influx of illicit 27MHz radio equipment from America and also the first country which was forced to capitulate in the face of widespread abuse. The Australian government actually established the precedents for what has since become a fairly traditional route for this sort of thing. First the pirates, on a small scale, followed by repression via due process of law. Next a huge upsurge in illegal use, leading to a reluctant admission that something needs to be done. The Australians countered by making legal a VHF facility which was scorned and ignored. Abuse on 27MHz grew rather than diminished. Reluctantly 27MHz CB was made legal, but with plans to phase it out, or at least run a VHF service alongside it. The hope here is that once users see the benefits to be gained from the superior VHF frequency they will change away from 27MHz of their own accord. In the meantime, however, laurels go to the pirates.

The pattern has been repeated in Holland, which now has 27MHz CB using an FM signal and, lately, France, which abandoned proposals for a 928 service in the face of widespread use of 27MHz sets illegally imported.

What is not so widely known is that the Americans have

their own pirates as well. Throughout all the arguments which have raged on the subject of CB in this country the pro-27MHz lobby has been the largest and perhaps the loudest voice. Although for a long while they had no clubs, no organisation, no political lobby, they were the not-so-silent majority who voted with their fingers on the transmit button. Throughout all the bitter wrangling their argument was relatively simple. They pointed to the American CB facility on 27MHz and observed as loudly as they could that although it may have drawbacks it does, indisputably, work. Since it does work, they reasoned, and since it is so widely accepted, and also since the hardware for their system already exists both abroad and in this country, where it is supposedly illegal, why not just shut up and get on with it?

On the face of it that seems to be a reasonably effective argument, but it is in fact rather heavily flawed in several places, perhaps most heavily from within.

To start with let's not forget that right at the beginning when the American idea of CB was first conceived it was allocated a VHF frequency, and the 27MHz service was only introduced when it became clear that existing technology was not capable of sustaining the VHF facility. 27MHz then was a second-rate service right from the very beginning.

It's also a frequency that was originally chosen over twenty years ago. At the time it was introduced many things which we now take for granted in the fields of electronics and communications were not possible. The same micro chip which has taken western civilisation along such a rapid path of development has also left the 27MHz band a long way behind. In the late fifties when CB first poked its head out of the cradle the whistles and pops which accompanied it were a part of everyday communication. We were, after all, only just past cats' whisker radio sets and telephones were by no means in every house in the land; those which did exist didn't actually work all that well, and nobody expected a perfect phone call every time. Under those circumstances the primitive sort of contact afforded by the early 27MHz sets was acceptable by comparison. Nowadays we live in a world run by satellite communication and get annoyed if a direct-dialled transatlantic phone call suffers from a small delay. We expect to be able to talk to someone on the other side of the world as clearly as if he or she were in the same room. For some strange reason the pro-27MHz lobby seem to want a sub-standard form of CB offering a level of excellence which they would never tolerate in their telephone. We have the ability to provide a high standard of contact via CB and there seems little point in settling for anything less.

It's not even as if there were some world standard for 27MHz CB in existence. There's not. Although there are more than sixty countries operating a CB facility based on the American service there is no such thing as a standard for these. America runs a maximum of four watts on 27MHz AM with forty channels plus upper and lower sideband. France has just opted for half the channel allocation with no Sideband and a maximum of 2 watts output. Germany is nearly the same as France, but Holland has chosen an FM signal. Although based on the same frequency band, very few of the 27MHz services are actually compatible with each other. The argument that we too should adopt 27MHz so as "not to be different" is thus patently ridiculous. Everybody is already different. And not only is everybody already different, but they're all trying to be even more different; they're all looking for alternative frequencies to 27MHz as quickly as they can go, and despite the World Administrative Radio Conference (WARC) and a million other committees of varying length and importance, hardly any of them are looking at the same thing.

There has for some time been a feeling that there should be some kind of world standard for CB, but opinions on exactly what form it should take have varied. Recently everyone seemed to be heading towards the UHF frequency of 928MHz, although both Australia and America have VHF services around 450MHz. In any case the World Standard as proposed was always designed to run alongside existing services rather than in place of them. However, in countries which didn't already have 27 the new standard would have been an ideal opportunity to lead the way in design and manufacture of equipment.

But by far the most obvious flaw in the pro-27 argument comes when considering the activities of those people already using it who are currently held up as a glowing example of successful CB at work—the Americans. They have their CB pirates as well as other countries, and instead of fighting for 27 to be introduced they're trying to make it go away or alter it to the point where it would no longer be recognisable.

To begin with let's consider the power restrictions on the American 27MHz setup. The FCC limit maximum power to four watts and very few mobile or base units use less than this, although many hand-held sets may only use two or three watts. With a properly mounted antenna a mobile four watt rig should, on clear ground with no obstructions like tall buildings, hills or trees, give a range of about 20 miles. About. This is absolutely perfect for the service as it was designed—short-range, two-way personal radio. However, in certain conditions freak signals can skip several hundred or even thousand miles, with the result that long-distance contact can be established for short periods. This sort of DX work has long been popular with radio hams, who are authorised to do it in the terms of their licences, and very quickly became popular with users of CB—much in the same way, perhaps, as keen car-owners like to make their car go as fast as possible. In fact DXing is expressly forbidden to CB users by the FCC, but then the speed limit never stopped the street-racers . . .

It's only a short step from wanting your car to go faster than others to tuning it to do so, and it's only a short step from working skip by accident or design to making sure that your rig has a better range than it ought to have by tuning or bolt-on power. With a low-frequency signal like 27MHz bolt-on power is very simply achieved and has large rewards of range in proportion to time, effort and money expended, as well as large range returns for smallish power increases.

All that is needed is the simplest form of linear amplifier which can simply be connected to the rig in the feedline as an extra box. The wiring connections are simple enough for a child to make and in fact making the amplifier itself is well within the bounds of possibility for the average DIY electrician. Naturally enough the addition of a linear amp—a set of boots, or a burner, in the jargon—is frowned upon by the FCC. In fact it's illegal. But that hasn't stopped America from taking up the device enthusiastically. It's not illegal to make or sell the things, they're only illegal when they're in use. Your average Yankee CB pirate can pick up a burner at his local radio store and be thrumming out huge amounts of power in a few moments.

Bearing in mind that the maximum output is supposed to be four watts, it's easy to see how pirates with burners ranging in output from fifty to two hundred and fifty watts could easily get to be unpopular round their home twenty. Also they can get Hawaii, Australia and some parts of Europe.

They'll be unpopular near their homes because a power output like that can easily overdrive a rig which is close by. The results need not necessarily be pyrotechnic, but they're usually pretty amazing. As a rule the power-happy are pretty careful how they use all those watts, in keeping with the peculiar code of conduct which seems to affect criminals, but they're not *always* that careful or lucky.

Neither are they alone in feeling that the 27MHz service as authorised could do with a certain amount of justifiable revision. They are joined at this point by the Outbanders, who have a whole different conception of just how the facility should be organised. Outbanders have been around for some time in the States, and probably originated among the few clairvoyants who were using CB before the boom took off in 73/74. Up until then they were members of a fairly small and reasonably exclusive club, but when the millions took to the air they were driven away by the inexpert, the inattentive and the idiots. That's their point of view, anyway.

They feel that since the early seventies the FCC has failed in its brief to police the CB band and maintain order, proper procedure and all the rest of the rules. As the early pirates in the UK were the first to go for Sideband when things took off over here, in order to get back to a fairly empty bit of air, so were the Outbanders probably the first to do so in America. They probably had a bit of peace there for a while.

Sideband is worth explaining a bit, because it's a very simple thing which confuses people easily. A radio signal can be drawn on paper just like a graph. It takes the form of a looping curve which travels above and below a central line. The distance it travels above the line is exactly the same as the distance below it. Likewise the distance from one crest to the next is exactly the same as the distance between the crest ahead and the one behind—that's the wavelength. When you press the transmit button on your rig that's the pattern which emerges. It's called the carrier wave. When you speak into

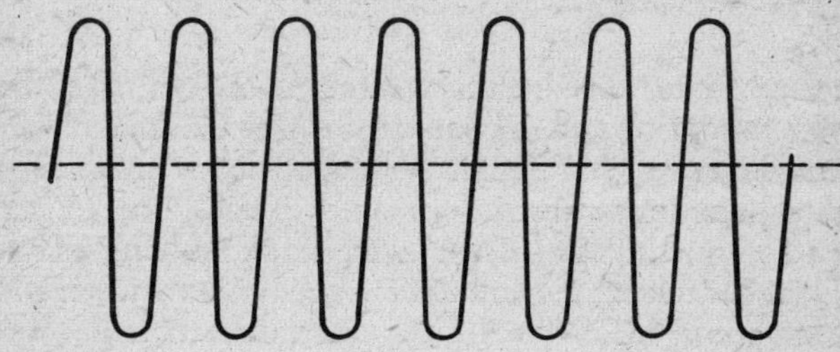

A simple radio wave generated by a radio set, without speech or any other form of modulation. The wave is symmetrical about a centre line and the distance between successive crests is the same.

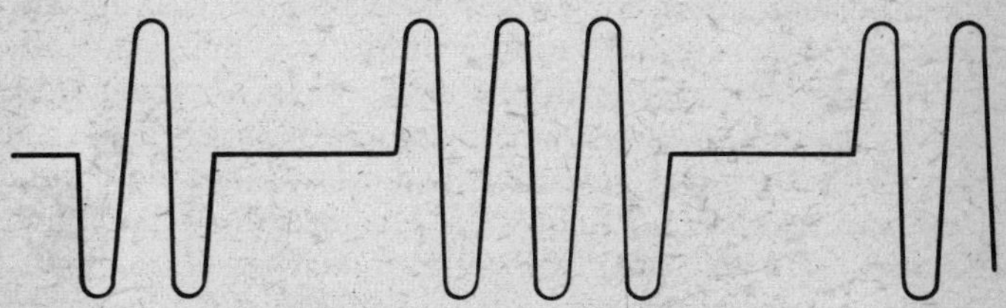

The same wave with simple modulation, like Morse or keying of the transmit button without speech.

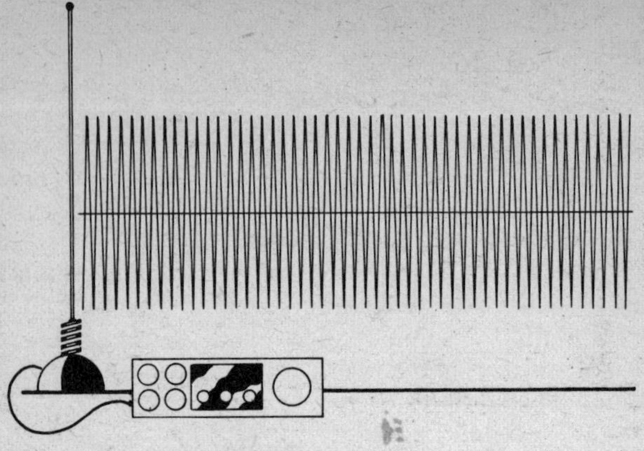

This is what it would look like if you could see it as the radio wave leaves the antenna; again this is unmodulated.

the microphone the signal is excited by the electrical whatnotery in your rig and the pattern changes in direct response to what you say. That's modulation, and the receiving set will unmodulate it and turn it back into speech. There are two types of modulation with which we might be concerned—Amplitude Modulation (AM) and Frequency Modulation (FM). With AM the signal distorts rather wildly and tends to leap higher and lower above its central line than the carrier wave. This clearly means that the signal is enlarged out of its bandwidth, and may bleed into the channels next to it if there is not enough distance separating them. FM signals are different, however. They stay the same size, but the waves get closer together or further apart at amazing speed. FM signals are obviously better than AM (ask yourself why stereo radio transmissions are always made in FM) because they stick to their frequency and can be placed closer together. You could get more FM channels in a given bit of airspace than AM. Also you will get much better clarity.

By a certain amount of electrical jiggery-wossname inside the rig it is possible for a radio set to transmit using only half of the channel—either those wobbly bits of the wave above

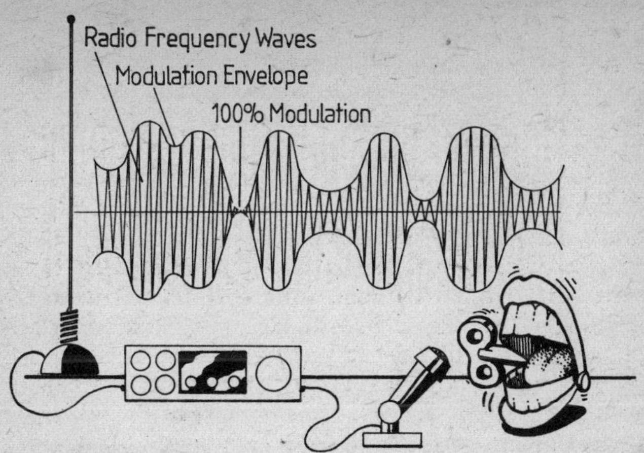

With speech added you can see how the radio frequency forms a word picture which is still symmetrical about the centre line; the receiving set only needs one half of the signal to be able to decode it. Using the upper or the lower half alone gives us Sideband.

the central line or those bits below it. This is Sideband and can quite clearly be divided into Upper and Lower Sideband. When using Sideband it is impossible for anyone who has not got a Sideband set switched to the correct portion of the band to hear your signal, although they may well get a load of mush through the speaker. In any event, using both Upper and Lower Sideband the number of available channels has just increased from forty to one hundred and twenty.

Sidebanders had quite a bit of peace for some time in the early days, and as the channels got busier so the convention which kept Sideband users in the high channels and ordinary AM users in the lower ones worked well for a while. But with forty-odd million people looking for a break all day and all night it's easy to see how those members of the early days private club might get a bit cheesed. Especially when they were of the opinion that many, if not all, of the newcomers didn't have a clue about radio procedure.

By acceptance 19 is the calling channel in the States. Find

the person you want, or a person you don't mind too much, then get off the calling channel to some other place and leave it free. Even when you've found a quiet channel you can't stay on the air more than four minutes and then you're obliged to take a minute out in case someone else wants to use it. Sadly this didn't happen all the time. Which is where the Outbanders came in.

Particularly in the early days of CB it was quite possible for a transmitter to wander slightly off channel. In those days there were only 23 of them, running from 26.965MHz to 27.255MHz, separated by 10KHz, with occasional larger gaps for model control and paging devices. Nowadays the picture's much the same except that there are 40 channels reaching up to 27.405MHz. But because of this wandering off frequency (which now does not happen, except on Sideband sets which have their own arrangements) it was possible to buy a Variable Frequency Oscillator (VFO or a "slider") which permitted slight tuning adjustments to the receiver side of a rig in order to pick up those stations which weren't quite on the spot.

It was not a long process to run the system back to front and deliberately tune both transmit and receive off the channel in order to create a brand new CB facility which was virtually the private club all over again. In time this has evolved to the point where the Outbanders operate on AM just below channel 1 and on Sideband just above channel 40.

1—26.965	12—27.105
2—26.975	13—27.115
3—26.985	14—27.125
4—27.005	15—27.135
5—27.015	16—27.155
6—27.025	17—27.165
7—27.035	18—27.175
8—27.055	19—27.185
9—27.065	20—27.205
10—27.075	21—27.215
11—27.085	22—27.225

23—27.255	32—27.325
24—27.235	33—27.335
25—27.245	34—27.345
26—27.265	35—27.355
27—27.275	36—27.365
28—27.285	37—27.375
29—27.295	38—27.385
30—27.305	39—27.395
31—27.315	40—27.405

This list shows the current channel spacing in use in America and consequently on all the 40-channel sets which have been brought into this country illegally and which were originally destined for the USA. In nearly all cases channel spacing is 10KHz except for occasional larger gaps which indicate the channels left clear for model control.

Even so there are many Outbanders now, whose numbers are measured in millions rather than hundreds. Not unnaturally the FCC takes a rather dim view of all this and does as much as possible to track them down and take them off the air. This has given rise to a strange paradox which was the pro-27 lobby's biggest drawback in this country. While British pirates are chased about by various forms of authority they maintain an honest front, pointing out that CB is a good thing, takes up little space, requires little or no effort on the part of authority to be made legal and in any case already exists, so why fight it? They further point out that in a democracy such as Britain the vast weight of public opinion is in favour of CB and governments who have denied the facility for so long are failing in their duty to provide the people with what they want—not what they ought to want. Or ought not to want. The "democratic freedom of speech" argument is probably the longest running and the one which generates the most emotion on either side.

As an illustration of Parliamentary attitude, during the House of Commons debate on the Marine Offences Bill (which put paid to the pirate music radio stations off the coast of Britain) one MP stated that he'd heard it said that giving in to public demand and allowing the pirates to continue was an indication of the democratic process at work. "It is not", he said. "It is pandering to populism". Perhaps it was a pun.

In any case the self-same argument currently rages in America. The Outbanders claim that the only reason for their existence is in the short-sightedness of the FCC, who failed to foresee the enormous demand that would eventually appear for CB, failed to allocate enough channels for it and then failed even to control the ones they'd got. Ergo: Outbanders. To begin with they were a jolly civilised bunch, policing their own private bit of air and even going so far as to publish their own set of rules. All of which works fairly well when there are only a couple of thousand of them in a country considerably bigger than the UK, but not so clever when there's three or four million of them. In short they've got the same kind of chaos on Outband as they have on the ordinary CB.

Meanwhile the FCC dispenses raids on Outbanders with the same kind of ferocity as did the Customs officers in England. Large fines and confiscation of equipment are commonplace; the FCC even monitor the ordinary CB bands and the amateur bands listening for operators who display a knowledge of Outbanding. Then they nip round and search for illegal equipment.

The Outbanders are righteously indignant. They believe that their cause is a just one, and that their service should be as legal as any other kind of CB. They organise marches, demos, even petitions to Government asking for their case to be considered and for the law to be changed. In their blacker moments they point out that America is the land of the free, and while it might not actually be the birthplace of democracy it's certainly the summer home. They intimate that no amount of stern government can prevent the American people from having what they want if enough of them want it, accuse the Government of ignoring its bounden duty and point to Prohibition as the classic case of this type. In short, the Outbanders are saying everything that the British pirates are saying, only with an American accent. And one other difference. They are already describing as useless the very thing which British pirates hold in reverence akin to that with which Monty Python regarded the Holy Grail.

Perhaps the most important fact which emerges from all this is the truth that despite the staggering size of the CB boom in America demand for two-way radio services of every kind is still growing. The most important lesson which the British government could learn from all this is that simply giving in to the pirate demands and legalising 27MHz CB won't be enough. They must also begin immediate planning for the expansion of the facility in as many directions as possible and to cater for as many users and types of users as will be needed.

The FCC have always pointed to GMRS as an alternative to 27MHz when talking about Outbanding, but even that is not sufficient, since it provides an entirely different type of service on a much-different frequency.

3

A question of choice—types of service, frequencies, the future

The arguments for and against CB have been heard so often (even in this book) that they don't really bear repetition. In any case that particular side of things seems to be long buried, and everyone has accepted that in the long run a CB facility brings more benefits to a community than it does drawbacks. All that's really left is what type of service do we require? What is expected from it? Who will use it? Why? What for? Once these questions are answered we can draw up a fairly general specification which will meet current and future need, rather than introduce a service which does neither and is never likely to. And by looking at other countries, principally America, since it has enjoyed the facility for so long, we can perhaps draw on experience already gained.

The Americans took only a short while to discover that 23 channels were insufficient, and even the expansion into Sideband only gave them 69, which were also insufficient. The advent of the 40-channel allocation, plus both Sidebands made a total of 120 which is now found to be not enough, and a further allocation is imminent. Obviously, and despite the gloomy predictions that a radio service would be undersubscribed, public demand actually increases in proportion to usage; that is to say that once it's started it becomes indispensable and gets used more and more often. The FCC have found this to be true in many ways. In particular many small (and large) businesses began using CB on construction sites and large warehouse lots and so on, only to find that it was so useful they wanted a better service with more accessibility and more privacy. Consequently the number of applications for private allocations (on different frequencies to CB) have expanded enormously. Cars were frowned on once —remember the red flag?—but who could live without one now?

Therefore we must be sure that whatever service we use will be capable of expansion to fill any growth which occurs. There are already a number of frequencies reserved for business use in the VHF bands between 70 and 90MHz and again between 110 and 130—low band and high band—but because demand is not that high yet sets are still quite expensive; their prices are measured in hundreds of pounds rather than tens. However, with the expected growth in demand this should change. What is unlikely to change is the amount of space available; if the upsurge of interest is proportional to that in the States then both bands together will not provide sufficient space. It is essential that such growth is provided for so that the CB band does not have to carry the overflow.

Likewise in America there have been large numbers of people who are dissatisfied with 27MHz CB on technical grounds and who are looking for a high-grade alternative. The GMRS provides such an alternative on a VHF frequency. By use of VHF and repeaters a much more effective service can be provided, offering less crowding and better range, albeit at increased cost. GMRS, however, is not CB in its true sense.

What we are looking for is a service which will offer short-range two-way radio to the ordinary citizen at the lowest possible price, with the highest quality possible and with the simplest possible means of operation. Radio enthusiasts, DX workers and DIY fiddlers must go elsewhere for their pleasures.

Writing the specifications for such a service is a reasonably academic occupation, since 27MHz CB is already here and will very likely never go away whatever happens. In addition there's nothing fundamentally wrong with 27MHz. It's just that technology available today enables us to provide something far superior in many ways.

However if you were to go looking for an area of airspace in which to place short-range radio communications you would end up somewhere between about 30MHz and 70MHz, probably nearer the higher figure than the lower. The trouble is that everybody else has already done it, and it's no accident that a great percentage of short-range military radio sets are tuned somewhere between those figures. Apart from any other considerations it is easier, as we have already pointed out, to transmit at the low frequencies than the higher, and far less power is called for. Thus a 2-watt portable on 30MHz will have a far greater range than a 2-watt portable on 300MHz. Bearing in mind that a CB service would be designed for base, mobile and portable use it's obvious that a lower frequency would achieve better results in return for power. In fact it's the portable set which is the limiting factor rather than either of the others, although the way the micro chip's going that won't be true for very long.

Low frequency transmissions are not all that wonderful, however. Below about 30 or 35MHz the wavelength is exactly right (or exactly wrong) for indulging in the phenomenon known as skip. All that happens is that the radio transmission goes on up into the sky until it reaches the ionosphere. Then, unlike higher wavelengths, it is reflected and bounces back to earth, often ending up thousands of miles from where it started. The propensity of the ionosphere to reflect skip is related also to Sunspot activity, which goes in 11-year cycles. 1981 will have seen a peak in the cycle and it will die off slowly, rising again to a high in about 1992. All of which is very grandiose in a small way, but it does mean that the cunning among us, perhaps aided by burners perhaps not, can get into some choice DX work if they pick their location. Conversation between Europe and America doesn't happen every day, but it's possible. However, it also drives legitimate users of the band out of their heads, since it prevents them using it for its design purpose—short-range communication.

Skip and weather conditions can affect any kind of radio transmissions in this way, and it's quite possible for a low-power PMR signal in the low band, designed for perhaps a thirty or forty-mile radius, to bounce from one end of Britain to another, but the lower the frequency the more likely it is to happen.

It is also true that all radio signals are capable of interfering with each other; how often have you had a taxi wipe out TV reception or the Gas Board walk all over your local radio station? In the latter case the two frequencies may not be that far apart—perhaps only two or three MHz, but *Crossroads* beams across the country at well over 600MHz and taxis are very likely in the low band—less than 90MHz.

Two factors are at work here. One is proximity. If the taxi or the Gas Board are right outside your house then their signal can easily swamp your receiver and blanket other incoming signals. You'll get picture fuzz across Meg Richardson, with perhaps a voice breakthrough and you'll definitely get crystal-clear voice breakthrough on the radio set.

Part two of this interference is not caused by proximity at all, but by harmonics. Nothing to do with music, harmonics are easy enough. First harmonic of 27MHz is 27MHz. Second harmonic is 54MHz (×2) and the third harmonic is,—this is a hard one—take away the number you first thought of and divide by five plus the number of corns on your left foot—81MHz (×3). Fourth harmonic is ×4, fifth ×5 and so on. It's easy really. The harmonic is a kind of shadow signal, not as strong as the original, rather like the "ghosting" effect you can see on badly-tuned TV sets. Exactly like it, in fact, except that it's on a different frequency. Anyone, therefore, operating a radio receiver of any kind on 54MHz is likely to suffer interference from the second harmonic of 27MHz CB transmissions. Which is why there was so much fuss in the States in the early days of CB, when American TV transmissions were roughly akin to our own Band One and slap in the path of CB interference.

The fitting of a high-pass filter to a TV set effectively cures all that anyway, and most modern TV sets should have such filtering systems built in. If they do suffer from interference it's because there's a fault in the rig or the TV, not because interference is a fact of life wherever CB is operated.

However it's easy to see how harmonics could get to be a nuisance. Especially if you start off realising that most of our radio transmissions take place from about 30MHz upwards—that includes Air Traffic Control, Radio rotten One, TV, all the Emergency services, all the PMR users, the works. There are a lot of 27MHz harmonics between 27 and, say, 900MHz, and you've clearly got more chance of hitting someone if you start off lower down.

Little snags like that can of course be cleared up, but remember that one of the first criteria should be that of price; the more technical masterpieces you build into the set the more it will cost, and if it's eventually priced off the market you've defeated the prime objective of a CB facility. So there has to be a balance. Even so, modern technology can produce an acceptably "clean" transmitter for a reasonable price.

VHF signals suffer less and less from all these problems, especially as they get higher, and on the face of it that does seem to make a VHF service more desirable. The first problem here would be one of allocation—definitely the PMR bands are too full already. And in terms of available radio spectrum Britain is probably the most overcrowded bit of air in the whole world—worse even than the States.

Actually finding a bit of space is a problem, although not insoluble, and the "Lancaster Band" is a prime example. In fact, 232MHz is ideally suited to the purpose. It's VHF, but not enough to make the price of equipment prohibitively expensive—prices will climb to an extent as frequency in use rises—while it does have all the advantages of VHF transmissions. Use of an FM signal would mean that there is plenty of room for about 100 channels, without taking up a lot of space and before we consider the addition of Sideband. The main drawback to this is found in the Outbanders. There is no doubt that a few, maybe more than a few, will soon find the existing system unsatisfactory for whatever reason, and will start to wander off channel. When they do the damage could be more than just a lost episode of a TV soap opera. Worse

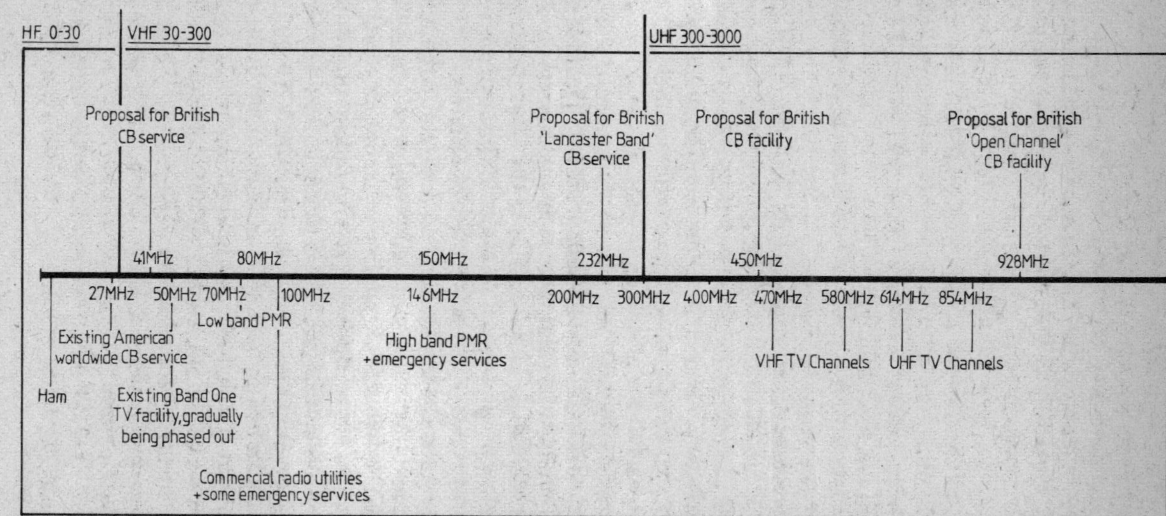

You can see from looking at this slightly out of scale representation that the airwaves have been getting a little crowded just lately. Even so there's enough room for at least one type of CB facility. Not shown, but spread throughout the HF and VHF bands, are the multitudes of channels used for communication by and with aircraft, shipping, even Diplomatic radio.

even than missing *Blankety-Blank*. There are all sorts of things happening around 232 which are a bit more important than TV or model control, or even hospital paging. VHF sets of this frequency would also respond fairly well to illegal amplification, although a reasonable range of, say, 20 miles could be obtained for a power output around 10 or 12 watts for the benefit of legitimate users. In many respects though, 232MHz could be said to be ideal—it has many advantages over lower frequencies and only a few of the drawbacks.

Going still higher, somewhere around 450MHz, roughly the same situation would exist. In fact this part of the spectrum is so well-suited to a personal radio service that it is the one chosen by the FCC originally in 1948 for GMRS. It would perhaps incur penalties of cost but would balance this by providing a very superior kind of service. 450 is probably too high for basic CB service use, and would be better suited for the next step up, perhaps as an overflow area when the PMR bands are full to capacity.

If we look any higher than 450 there are vast tracts of spectrum which are full up speeding assorted TV programmes around the country, and there is no real free space until we get over 900MHz. This is well into the UHF band and starts to have a whole new set of problems all its own. The two major areas for concern here would be cost, which would be well above the maximum level that would make a CB facility worthy of its name, and the second would be range. Even with a 25 watt output it's unlikely that mobile to mobile communication, across open ground, would be better than three miles. This figure is so low as to make the frequency useless for a practical CB facility. And since it is so high, it would not repay the investment of extra power output at anything like the rate of lower frequencies. With 50 watts mobiles might extend their range to six miles over open ground, but a crowded urban environment would reduce the UHF line-of-sight signal to travelling as far as the driver of a car could see and no further. Morse Code with the headlights would be more effective. Furthermore a 50 watt set, especially as a hand-held, would be expensive, bulky and possibly dangerous to human life as a result of non-ionising radiation from the antenna. UHF frequencies have recently found much favour for expansion (or in some countries introduction) of in-car telephone services, and to this purpose they are reasonably well suited, especially as subscribers to such a service expect a high cost to retain the exclusivity on which it would depend for successful operation.

But for CB we are definitely confined to something appreciably lower than 450MHz. Spectrum crowding dictates that we choose between 232 or something below 50MHz, and in this the existing 27MHz service seems to be the front-runner on the same basis that Hilary chose to climb Everest rather than build his own mountain—because it's there. The hardware exists and the teething problems which the creation of a new service like this is bound to encounter have already been met and dealt with.

The only real alternative to this would be the old Band One TV frequencies just above 40MHz, to which existing sets could be modified easily and cheaply, as could the production equipment currently engaged on the manufacture of 27MHz sets. This should please everybody in the electronics industry who isn't British.

The biggest problem here is that opting for 27 doesn't take into account the possible future developments of personal radio. The cordless telephone is already here; the cordless CB microphone, utilising exactly the same principle is also here. CB phone patches are a reality in the US. Selective-call CB sets are now a reality, thanks to the micro chip. CB sets can now be made to "ring" in response to a personal number code just like a phone. All you need now is to be within about one mile of your home or your car in order to receive a personal call which may come to you via the CB network or the telephone service. American thinking is now heading towards the time when the CB network and the phone service will combine and become the same thing. When they do, the direct-dial, person-to-person transatlantic telephone call via your wristwatch will be only a step away. The technology exists, the legislation doesn't. But by 1985 . . .

The problem here is that, if we use 27MHz for CB over here like everywhere else, such a sophisticated service as this wouldn't be on that frequency, because it cannot offer the required quality and integrity. Something like that is going to have to be VHF or it won't work. Of course it could, and probably will, run alongside the existing 27MHz service rather than replace it, which makes the whole process of arguments like this pointless.

In the end we're left with a requirement for a service which provides two-way radio communication as cheaply as possible with the minimum of formality and the minimum requirement for specialised or technical knowledge without causing interference to other users. With the possible exception of occasional interference to current hospital paging systems it would seem that 27MHz is that service.

4

Choosing a rig—faceplate controls and their functions

Let us now proceed to that happy state of grace whereby you are about to walk into a shop and buy a rig. Possibly several, depending on your requirements. Or on the number of cars you own. The frequency in question matters little at this point, since basic "black box" radio equipment looks the same and is operated the same way regardless of frequency.

Further standardisation will be forced upon us by virtue of Government control. All rigs are limited to a maximum output, and as long as you don't buy one lower than maximum (and it's unlikely that such a think will exist, except perhaps as a hand-held item) then there won't be much difference between the sets on sale in the shops. The component specifications will be held to a minimum level as well, so that even if you buy the cheapest mobile set you can find it will still perform to a given standard. Bleed and crosstalk will not be allowed to fall below a certain standard, for example, and there is little likelihood of manufacturers voluntarily spending their money on improving this sort of thing. There will always be the upper end of the market for enthusiasts, of course, but these sets will be recognisable easily enough by their price, and needn't concern us at a basic level.

With so much standardisation creeping in there are a limited number of ways in which a manufacturer can attract potential purchasers to his product. The emphasis in the States seems to have settled on gadgetry and appearance, plus a fairly high variety of superfluous nonsense (like digital clocks and so on) which is currently afflicting the in-car-entertainment industry. It will be necessary for you to be able to distinguish between tinsel and function though, and while we may all be able to see that a digital clock is unlikely to have a noticeably beneficial effect on reception there are other items which may confuse or mislead.

To a large extent purchasers have dug their own hole in this, because everybody likes to show off. The maxim "if you've got it—flaunt it" applies as much to electronic hardware as it ever did to Norma Jean. It's exactly the same kind of gadget snobbery which is so prevalent in the domestic and mobile audio market. There are very few purists who actually want (or can afford) a hi-fi setup which will allow them to hear Mario Lanza's truss straining during a solo, but anyone who has got equipment which can do that automatically requires the equipment to have the appearance of being able to do so. By exactly the same token people who have equipment which isn't capable of such refined sensitivity also require it to look as if it is. Which is where so many of the useless knobs, dials and flickering lights come into the story.

Now it's a fact that in the early days of CB many of the sets, which we might now be forgiven for regarding as being slightly primitive, required some extra help in getting things right. Many of them therefore had a confusing number of controls, all of which contributed to their performance. But as the "chips with everything" brigade got going the picture changed. Firstly the sets became sophisticated enough for a number of these controls to become unnecessary, and secondly it was possible to incorporate some of the others into the circuitry so that they operated automatically, leaving the face of the unit with a basic minimum of controls. Motorola were probably leaders in this sort of thing, but they soon found out that it didn't pay off. The sets may have been as good as, or even better than, their rivals. Certainly they were simpler in operation. But they didn't look as if you could speak directly to Mission Control. You couldn't. Any more than you could with the gadget-ridden competition. But the competition *looked* as if you could and that counted for a lot when it came to paying out money.

Liken it to a car. Early vehicles had a magneto advance/

retard lever, usually on the steering wheel, which altered the ignition timing for starting, accelerating and cruising. Then some bright spark produced automatic vacuum advance/retard and the mag lever got binned quicker than yesterday's news. Now we're asking for it back. Motorola have had to give it back. Dials, switches and gadgets are very definitely in. But you get nothing for nothing, so it's worth knowing what they're all for so that you can at least decide how much snobbery you're willing to pay for.

On/Off/Volume
Piece of wossname, this. Turn it till it clicks—the set's switched on. Keep turning it and it gets louder. Just like your car radio/TV set/hearing aid. It may be worth mentioning that this only affects the volume of incoming signals. It does not boost their strength or improve reception, neither does it affect the volume of your transmissions.

Channel Selector
Exactly that. If it says 20 you're on channel 20, if it says 16 you're on channel 16. If it says 48 it's lying to you or you've broken it. This may come in the form of a mechanical wheel with an illuminated window indicating the channel in use, a wheel with an LCD display indicating channel in use or simply as an LCD display with two buttons marked "up" and "down" respectively. If you're on channel 12 and you want 24 press "up". If you're on 12 and want 6 press "down". Pressing "up" from 40 gets you 1, "down" gets 39. This latter type very often revert to channel 1 every time they're switched off, which can get to be a pain after a while.

Squelch
A lovely name, squelch, and somehow remarkably apt in view of the function. It is a fact that the airwaves are full of a lot of noise. Listen to your tranny after Radio One goes off the air and you'll hear all manner of whistling and scratching. "Mush", "hash", "static", call it what you like, it's all a bloody nuisance. Your squelch control has thoughtfully been provided so that you can reduce all this to a practical

minimum. With the rig switched on and the volume at an acceptable level, set the squelch fully open. Now gradually close it (anti-clockwise is a good bet) and you'll hear the level of background noise slowly die away. Before the squelch is completely closed, and usually about halfway, it should die away altogether. At that point the set will respond to transmissions which are close enough or strong enough to give an intelligible signal but won't drive you out of your head the rest of the time. The sensitivity of the set is somewhat reduced by this, and you will be able to maintain contact with a distant station moving away from you longer by opening the squelch, but the penalty will be increased background rubbish. Eventually the distant station will be blotted out by background which means you're out of range or he's gone home for tea.

Transmit or Push-to-talk (PTT)
This you will find on the microphone, usually in the form of a bar pivoted at one end. Push it down when you want to say something and release it when you've finished if you ever want to hear anything.

Automatic Noise Limiter/Noise Blanker (ANL/NB)
A two-position switch which is just like the scratch filter or noise filter on your hi-fi set. The Dolby of the CB world, in fact. A good one won't affect the sensitivity of your set as such in the way that squelch does, although it will have much the same effect. What it does is cut off the top audio frequencies of incoming signals, since these are the ones which carry the bulk of the interference. Within working range it will give you a clearer incoming signal free from crackle. Outside that range you're stuck with the noise.

Distant/Local
A two-position squelch, almost. Talking to people very close to you can overdrive your set. Switching into "local" will prevent this, also remove distortion. It also cuts down on background noise since it reduces the sensitivity of the set and eliminates weaker signals. Switch to "distant" to regain

full sensitivity and contact with those far-away breakers.

Power Meter and Signal Strength Meter (S-meter)

The power meter shows on a scale of, say, 1–5 what proportion of available power you're putting out. The nearer the top it is the better you're doing. If it's at the bottom you're not.

The S-meter shows you the strength of incoming signals on a scale of, say, 1–9. If it shows 1 you probably can't hear anything anyway and if it shows 9 you're about to go deaf. Hence "pegging the meter" (a full-strength reception) and "give me a 9" (say that again). The S-meter is unrelated to volume, so if the meter shows 9 and you can't hear a thing turn it up.

Mode Selector (AM/USB/LSB)

Only three positions, and only applicable to sets equipped with Sideband capability. AM is straightforward, normal operating. LSB is Lower Sideband and USB is, surprisingly, Upper Sideband. Really couldn't be simpler, could it?

Delta Tune/Clarifier

This little number will enable you to fine tune the receive side of the rig only, in order to zero in on transmissions which are slightly off the edge of the channel for some reason. Anyone who is on the wander should be told fairly smartly. In practice this control will only be needed if you're operating Sideband, when it can make quite a difference, and when a slight amount of straying is permissible if not actually expected.

PA/CB

Most rigs seem to have this capability nowadays. Plug in an external speaker to the jack provided and switch to "PA". Your rig stops being the CB unit it was in the "CB" mode and becomes a loudhailer, responsive to the transmit button. Theoretically this should be very useful for things like crowd control at the scene of an accident or similar, but in practice it gives much pleasure when used to imitate the voice of the Almighty for the benefit of unwary drunks, or to explain to the driver in front exactly why you made that gesture and exactly what you're going to do to him at the next set of lights. It can get you into a lot of trouble and really shouldn't be used unless you are a greater deal bigger than most people you know.

Mike Gain

This is the only control which will affect either the quality or volume of your outgoing transmission. All the rest can only alter received signals. It affects microphone volume and is a refinement which means that you can hold the mike where you find it's comfortable and adjust the gain to suit. There is a point at which your transmitted speech will begin to distort and you should enlist the aid of a fellow-breaker to find the optimum setting and leave it there until you get a bad cold or tonsilitis or something.

That lot wraps it up as far as the basic cooking rigs are concerned. There are many more controls available, especially with the newer micro chip stuff beginning to arrive on the market, but all that is just trimming, and the rig won't work any better at its basic function than before. Of course it will do a whole variety of stuff which was never even dreamed of even a couple of years ago, some of which you may be willing to pay for. Or not.

We've already mentioned selective calling, and these rigs, when in "selective" mode, can only be activated by an operator of a similarly equipped set who knows your private number.

Other rigs may use a tone squelch which is much simpler and responds only to a tone—this is clearly less selective than a private number. Some rigs will scan for a busy channel, some for an empty channel, some for either. Some constantly monitor channel 9 and switch over to it automatically if it gets busy. Others will switch to a pre-determined channel at a selected time, or just switch on at a selected time. A few will even monitor their own SWR and display it if asked to do so; they will also switch off if a fault develops and the SWR is too high, alerting the operator by flashing a dial full of zeros on and off. None of them can make apple pie . . .

feedline which need intrude into the passenger compartment of the car will be the one to the mike itself, but this will be carrying all the command functions as well. Consequently it will be very thick and have an absolutely giant plug on the end of it. Naturally this will require a giant hole in the bulkhead. When you've found the spot, free from vital impedimenta, make sure you can get an electric drill with a hole-cutter to it. If not—find another place. You'll never make a hole in a car bulkhead with a toothpick.

Now, if it's all planned out, and all the wires reach without stretching, chafing or kinking, the set has a firm mounting within sight and easy reach of the driver, and it doesn't interfere with the operation of the vehicle controls—you can go ahead and screw it in.

As a last thought, bear in mind that stealing cars is one of the fastest-growing participation sports in the country and stealing from cars is the next best thing. Joe Criminal will only nick stuff if he knows it's there, and then only if it's jolly easy. You can save yourself a lot of bother by taking some elementary precautions. First make sure that your rig is covered by insurance—some policies don't cover accessories, or only if the rest of the car is stolen and found minus the expensive bits, so make sure.

Making your rig as unobtrusive as possible is also a good idea, but sadly it usually conflicts with the need to have it where it can be operated easily. Using a slide-mount is quite smart, as you can whip the unit out in a second and put it in the boot or take it with you; replacing it also only takes a moment.

Make the fixing as strong as possible—if stealing takes time the blagger has more chance of being caught and is likely to give up or not start at all. In the end nothing short of welding the unit to the chassis will stop a determined thief, and even then you'll only encourage him to take the whole car to somewhere he can work in peace. Investing in a burglar alarm isn't a bad idea—it only costs a few pounds and can save you a great deal of distress. And if you can install a rig you can certainly fit an alarm.

5

Mobile antennas—choice, installation and SWR

Now that the rig is in place you'll want to start using it. Don't. Being a sensitive little object, your rig does not take kindly to being switched on without first being connected to an antenna. It will not take kindly to being used for transmitting purposes unless that antenna is correctly matched to the set. In fact all that electronic wizardry is a load of useless garbage without a simple stick of metal. Although it's perhaps not that simple.

We already know that radio signals travel in waves, with a regular spacing called wavelength. In all cases the best performance is obtained by using an antenna which is the same size as the wavelength. Sadly, in the case of 27MHz, this involves a mast in excess of 30 feet in height, which is clearly not a practical proposition, especially for a car or small truck. Luckily the next best thing is an antenna which bears a mathematically proportionate relationship to the wavelength—half or a quarter of the size, for example. Easily enough these are described in the same way—half-wave, quarter-wave, five-eighths-wave and so on.

More bad luck comes when you realise that even a quarter-wave antenna will be roughly 9 feet long; this is acceptable in some cases, particularly for trucks, but is still not terribly practical. Like in multi-storey car parks, for example. All of which would present something of a problem were it not for one saving factor. Antennas, as a general rule, are not overburdened with intelligence and may be fooled into believing that they are longer than is actually true. In fact it is the electrical length of the antenna which is important, rather than physical height. So if you accept the optimum height of the antenna as 9 feet it is possible to make one 3 feet high and add to it a piece of wire six feet in length which, when coiled tightly, is only four inches high. This is called a loading coil, and produces a mast slightly more than 3 feet tall, with an electrical length of exactly the desired proportion. This subterfuge is widely practised and, though effective, is not as good as the 9 foot mast. Positioning of the loading coil can also affect the performance somewhat and thus you will find antennas with the coil at the bottom (base-loaded), halfway up (centred-loaded) and right at the top (guess). Since the length of the antenna is so important they are also provided with a means of adjustment. This is easier than most people think it is and is the last thing you will need to do to your antenna.

Before that you will have to make a number of decisions. The most obvious is mounting. Where will you put it? Next, how will you fix it there? These two are very much related, and it's worth making both decisions together.

Because your car contains large amounts of metal it will affect the radiation of your CB signal. Metal-bodied cars will do this more than GRP types, but then the GRP owners are going to have real problems with suppression later. The bodywork will in effect mask your transmissions to a degree, and the effect of this on signal pattern depends on where the antenna is placed. As a matter of practicality it probably won't bother you that much; it's obvious than an antenna on the roof will do better than other kinds, but after that it's fairly academic. You won't know in advance which side of your car you'll be speaking to, so the choice is impossible to make. Interested parties can purchase little devices which are sensitive to the radio output of the rig. As you walk round the car with the set on transmit it will buzz when it gets a signal and won't when it doesn't. In this way you will know where and to what extent masking is taking place.

Since antenna location is traditionally achieved by boring large holes in the bodywork of the car into which the mast is bolted you may find that your choice of mounting is dictated to a large extent by the amount of free space on the bodywork available for the purpose. Before you drill any holes make sure there's nothing underneath to get damaged, and make sure that there's also enough space beneath the panel to accommodate the undercarriage of the antenna. This latter won't generally be a problem unless the loading coil is beneath the base of the mast (unusual) or unless the mast itself is retractable (becoming more and more common). Make sure also that the feedline can be led to the rig without kinking or stretching or chafing. If it's not long enough then extension leads are widely available. Do use proper, soldered connectors for joining the leads up. It's worth mentioning that very many of the retractable types of antenna are designed to double up for a CB and regular car radio use. This will save a lot of messing about and will also make the appearance of the vehicle very much neater. It's another way of fooling thieves as well, which you may wish to take advantage of.

Don't leave the feedline loose inside the car or dangling underneath where it can be damaged or severed. Damage will hinder the performance of the rig while severing the cable will destroy it.

If it all checks out and you are certain that the installation is okay then you can drill your mounting hole. Clean the bodywork around the edges of the hole right back to bare metal so that the antenna gets a good earth, otherwise you'll have trouble from interference later, and possibly poor performance also. It may be that you don't relish the idea of making holes in your car, or it may be that you change cars frequently, in which case you might opt for one of the variety of alternative mounting methods. Broadly these fall into two categories—those which clamp on and those which stick on.

Seen from above the different radiation patterns afforded by different locations of the antenna (signified by the black blob) on the car body are clearly not the same. In practical terms, however, you won't be able to tell them apart once you're sitting in the hot seat with your finger on the button.

Clearly neither of these types will have the quality of earth connection enjoyed by the traditional antenna, but they have a flexibility which may or may not compensate—it's a question of personal need and preference. The clamp type will clamp almost anywhere. As a guide they are available for clamping to bumpers, bootlids and rain gutters. They are semi-permanent devices, which makes them semi-portable; they can be removed easily enough at any time, and just as easily replaced.

Less permanent, and even more portable, are the stick-on type. With the exception of the models which may be glued to the window of the vehicle these are simply stuck to the car by means of a giant magnet, and are thus called mag-mounts. And though it does not require the strength of Geoff Capes to remove a mag-mount no amount of fast driving will blow them away, although people of a suspicious nature will want to keep checking in the beginning. The advantage of the mag-mount is that you can put it anywhere you like. Most seem to end up in the centre of the roof since that is the best place for performance. And although fixed masts of any kind are susceptible to damage by vandals, whether they be the teenage variety or the automatic carwash kind, mag-mounts disappear into the car in a trice and may be replaced with equal facility. This will also discourage thieves, who are less likely to steal your rig if they don't know you've got one. Coupled with a slide-mounted rig this kind of antenna can be especially attractive to owners of soft-top cars.

The biggest problem affecting clamp or mag-mount antennas will be that of the feedline. The simplest method is to run the line in through a window, but sooner or later, especially in British weather, you or a passenger are going to be tempted to shut the window as tightly as possible, with distressing effect on the coax (co-axial cable). It is often possible to run the line round door edges, but you will have to be careful when opening and closing the door, especially if you demount the antenna frequently and leave the wire loose. If you don't, then broad plastic sticky tape will hold it in place for an appreciable period, but the problem is always one that you'll have to allow for. Once again it's a matter for personal choice.

Once the antenna is in place and properly connected it's time to switch on and carry out the fine tuning of its length. For this you will need patience, the little adjusting tools which came with the mast and an SWR meter.

SWR only means Standing Wave Ratio, and is not the mystic formula which most people seem to think it is. In fact a great percentage of American CB users don't even know what it is; of those who do, not many actually care. This is probably due to the high number of factory-fitted or dealership-fitted rigs on the roads over there, but even though it makes a difference to your rig it is likely that poor connections or damaged cable will have more effect on SWR than the antenna trimming itself.

Without getting involved in the complexities, an SWR reading indicates how closely your antenna length is related to the wavelength you are using. The ideal ratio is 1:1, but that's found about as often as hen's teeth. All you can do is get as close to that figure as possible; anything under 2:1 is acceptable and unless you have the manual dexterity of Fagin coupled with the patience of Doctor Kildare you'll never improve on 1.5:1.

In order to adjust your SWR you will need an SWR meter. Like everything else connected with CB they come fancy or cheap, and the cheap ones will do the trick well enough. You can buy a meter at any CB shop and at the time you are well advised to purchase a patch lead as well, otherwise you won't be able to use the meter. There will be two connections on the meter, one of which will definitely be marked "ANT" and the other will probably be marked "CB". The front will have a numbered dial, a knob like a volume control and a two-position switch marked "forward" and "reflected" or abbreviations of same.

Armed with this device, approach your car. Park it in some clear space—not under trees or next to your house or garage. While the rig is switched off disconnect the antenna feedline and screw it to the terminal on the meter marked "ANT".

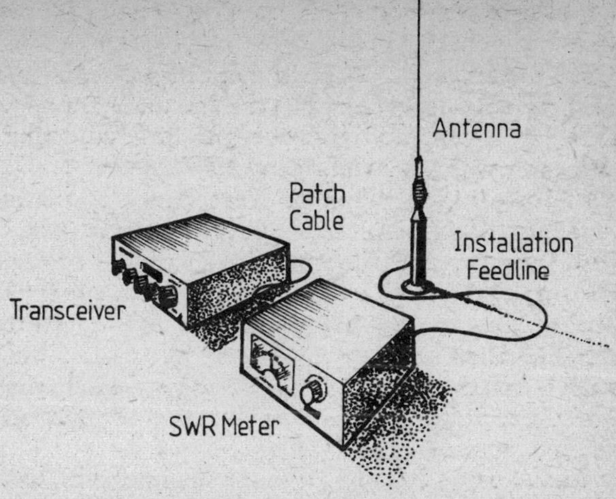

If you've been doing everything right it should all run in a straight line from antenna to rig and you should be able to sort your SWR out with less bother than the Artful Dodger stealing pocket watches.

Using the short patch lead which you so wisely acquired earlier, connect the meter terminal marked "CB" or whatever to the rig antenna plug. The meter is now in the feedline between the rig and the antenna and the hard part is over; from now on it's all downhill.

Make sure that all the car doors are shut, plus the bonnet and boot; these metal objects can have a slight, but measurable, effect and since you won't be driving around with them flapping it's preferable to set your SWR with them closed. Now switch on the set and observe that nothing has changed, nor is there any smoke etc. Now, since changing channels changes frequency (namely, wavelength) it follows that a perfect antenna for channel 1 will be of a minimally different length to a perfect antenna for channel 40. Not that different, and since we've already discovered you're not blessed with the manual dexterity of Fagin etc, you won't be able to do very much about it anyway. But since we're doing everything properly we'll account for it. Basically you've got a choice.

Either select a channel in the middle of the dial—19, 20 or similar—which will give you blanket coverage, or choose one you use most often—14 or 19 are common, 9 if you're accident-prone.

The next thing is to switch the meter into "forward"

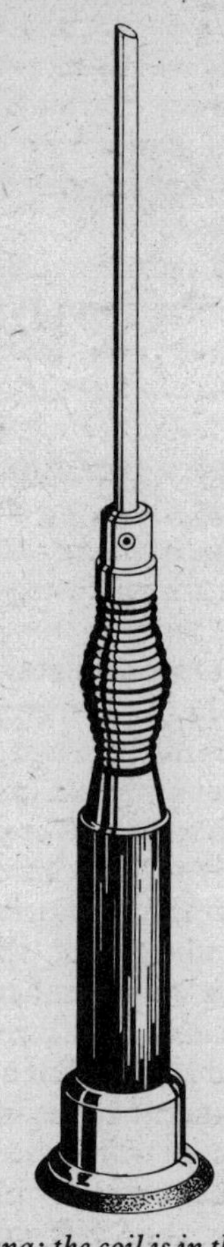

Typical base-loaded antenna; the coil is in the black base and the adjustment for antenna length (achieved by means of an Allen key in this case) is just above the spring base.

mode. Then press and keep pressing the transmit button of your rig. You will notice at once that the needle in the dial of the meter has moved and now indicates some figure. Ignore it. Twiddle the knob like a volume control and the needle moves some more. Twiddle it so that the needle is giving its maximum reading. Then switch the meter to "reflected". With any luck the needle will drop to the bottom of the dial and give a figure somewhere near or below 2:1. If it's below it, stop worrying. If it's above it or you're a perfectionist blessed with the manual dexterity etc then it's time to make some adjustments.

If you have a dual antenna for CB and music radio the feedline will divide at some point through a junction box which is properly known as a splitter box. If this is the case it will have a trimming screw for the CB which should be adjusted with an SWR meter before the antenna length is adjusted. Use the same procedure for tuning the trimmer as for the antenna but instead of making it longer or shorter turn the trimming screw clockwise or anti-clockwise. Once you've found the best setting for the trimmer you can then adjust the mast length.

If you examine your antenna you will find that somewhere there is provision for adjusting its length. As a rule retractable jobs are fractionally too short and have a threaded tip which may be extended and locked into place. Simple whip antennas are made fractionally too long as a rule and are thin enough to have the tip shortened with wire cutters. Generally they are held in place with a locking screw at the base and there is room for them to be moved up or down without the need for cutting. In any case don't cut anything yet, and when you do remember that you need only remove a tiny piece of the mast to make a big difference to the SWR—once you've cut off too much you won't be able to stick it back on.

There is a sort of short cut which can tell you whether your mast is too long or too short. Simply put your hand near the mast and observe whether the SWR goes up or down. This is not always as easy as it sounds because there's a fair chance that you won't be able to reach the antenna and watch the meter and press the transmit button of your rig all at once. If you can, and you haven't got the car door open, I'd like to meet your sister. It's possible that you may be among friends when performing your SWR check, in which case get one of them to put an arm near the mast while you watch to see if the SWR goes up or down. If it goes up then the mast needs to be shorter and if it goes down then it needs to be longer.

Alternatively, and especially if you haven't got a friend, after you've taken the initial reading, adjust the mast up or down—it doesn't matter which—and then check the reading again. If it's gone up you were going the wrong way and if it's gone down you were going the right way. Keep making tiny adjustments until you're below 2:1. Remember that there is an optimum length and if you pass it the reading will start to climb again. Whips which are still too long, once they are seated at the bottom of their sockets will have to be trimmed. Minutely, right?

When everything is as good as you can make it, switch off, remove the meter and reconnect the antenna feedline directly to the rig. Now you can zoom off into the sunset. In all probability your ears will be full of the most painful crackling imaginable, all of it emanating from the rig. Engine interference. You, like so many of your kind, need suppressing.

6

Troubleshooting—suppression, interference and TVI

We have already established that the air is full of all kinds of unnecessary radio signals which, when received by your rig, can best be described as interference. In most cases the suppression of all this rubbish is handled as well as it can be done by various filtering methods built in to the rig and which operate either automatically or by means of one of the various controls we've already discussed. All that you need to do is to make sure that your rig is installed properly in order to make these gadgets function properly.

Unfortunately for the mobile breaker the greatest source of radio interference is likely to be the car itself. Modern legislation applies fairly stringent requirements concerned with the minimising of such interference to vehicle manufacturers, but in most cases these are designed to protect domestic or other TV and radio devices not mounted in the car itself, and are not sufficient for our purposes.

The first way of eliminating interference is to ensure that all connections to the rig are clean and firm, with particular emphasis on earthing. This applies particularly to the antenna. If the antenna feedline is damaged in any way then the performance of the rig will be affected in two ways. Firstly it will collect as much interference as available and transport it directly to your speaker, and secondly it will hinder the broadcasting of your signal. Effectively, once the outer screening of the coax feed is damaged it ceases to function. Since the coax is used to screen the antenna feed you will experience this incoming interference and the feedline itself will begin to act like a giant antenna. Since the length of the antenna is so important it is easy to see that your signal strength will diminish and your SWR will rise to a point where you'd be better off switching to PA and shouting as loud as you can. Damaged cable, or worse still an antenna feedline which is not of the coax type, is a definite no-no.

Assuming that all this is as it should be you will find that the bulk of your interference only occurs while the car engine is running and/or certain pieces of its electrical establishment are functioning. The trick here is to pin down as accurately as possible which things are making the racket, which is not actually that difficult.

Switch the rig on and the engine off. Then operate everything electric—heater, wipers, washers, flashers, headlamps and sidelights (lights won't interfere themselves, but their operating relays may well make a definite click as they are switched on and off)—one item at a time. Make a note of those items which make a row—the heater, washers and wipers are most common. If it's everything then your problem is basic—go back and check the installation again. If it still persists go on to the next step.

It is easy enough to buy suppression kits for your car, very similar to the type which you might fit in conjunction with a normal music radio. You can buy the same sort of thing for your CB, although to a somewhat higher specification. In practice suppressors are only capacitors anyway, which electric current recognises as a discharged battery and rushes

Suppressors. They all look much the same when they're of this general-purpose type for silencing the irritating crackle of on-board accessories. Some are more powerful than others, but you can't tell by looking.

into, rather than rushing anywhere else, such as into your CB.

It can be a pain fitting individual suppressors to every item on the car which is interfering with your rig, especially if there are a lot of them, and there are one or two short cuts worth trying, which may also go some way to eliminating ignition noise which is the loudest and nastiest interference. That won't arrive until the engine's running though, and we haven't got that far yet, so pay attention.

If both power and earth leads of your rig are not already connected directly to the battery you may wish to alter the situation so that they are. This can often remove slight interference from the vehicle components. If it doesn't, or if the rig was already so connected the next easiest thing is to install a hotline filter in the power feed. This is a broad-spectrum suppressor of considerable power and should eliminate a large amount of racket. If not, you'll have to go round every damn thing in turn, bolting on loads of individual suppressors. Wearing, and cumulatively expensive.

But still, with the engine off, the airwaves should be reasonably peaceful after it's all been done and we may safely proceed to the next step—switching on the engine. This will

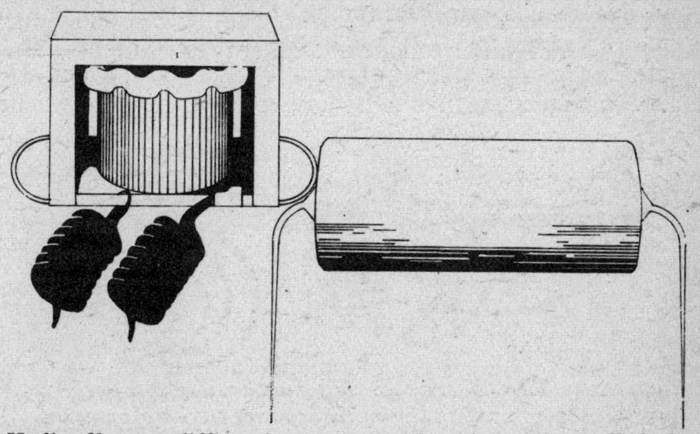

Hotline filters are different in appearance altogether, and actually look as if they're more powerful and designed for a bigger job. They are.

indubitably provoke a cacophony of noise not totally dissimilar to that available by tuning your music radio to 275 metres in the medium waveband. It is far louder than any other kind, harder to track down to source and harder to eliminate. But it can be done by following a simple rule: trust me.

Next—lisen to the interference. If you can put up with the strain long enough it will eventually separate into at least two distinctive types—one of them a pronounced and vicious crackling and the other a whining noise. Both of these will rise in pitch and ferocity as engine revs climb. The whining noise will be the generator, and if you're lucky it may not actually occur at all. The source of the whine may be verified by the not-so-simple expedient of running the engine without the fanbelt, which involves the undoing of several well-hidden and seized bolts. Haemophiliacs are advised not to attempt this unsupervised, as engineering activities of this nature will infallibly make you bleed.

Fitting a suppressor to the generator is simple enough and follows the same procedure as individual suppression of the other electrical components of the vehicle—bolt the body of the suppressor to a good earth and connect the lead (there's only one) to the live wire of the object in question. This should produce a gentle hush. Smashing.

Next—the ignition, and this is where it gets naughty. The high-voltage electricity necessary to make your engine work

This is the sort of low-powered job you would use to suppress your ignition and/or generator if you were installing an ordinary music radio. The chances are you've already got one. The chances are also that it won't be much good for your CB, although you could get lucky. Then again . . .

is the loudest and most common form of interference and will come from three sources. These are the spark plugs, the ignition coil and the distributor. Basic rules apply first, and may save time, trouble and expense later. Make sure that the antenna feedline doesn't run close to any of these devices, and make sure also that the earth or power lines to the rig itself don't either. Doing this may eliminate a certain amount of crosstalk and make life more peaceful. It's unlikely that this alone will clear the whole trouble up, however, and more will need to be done. The easiest and cheapest thing is the fitting of an individual suppressor to the coil but this still is unlikely to provide you with the final solution. Make sure it's fitted to the lead between the coil and ignition, as this is the one which carries all the current making the crackling noise.

Ensure that all the HT leads—between coil, distributor and plugs—are either of the suppressed carbon-fibre type or, if copper-core, are individually suppressed. Make sure that the plug caps are of the suppressed variety also. In theory doing all this will make all the racket go away. If it hasn't gone there are still a few tricks left in the armoury. Most of them relate to earthing of various components. Antenna earth we've already stressed several times, but you may have to go to fairly extreme lengths with this. Before that, check and clean, and replace if necessary the battery earth strap and

HT suppressors are built to take care of the fiercer, high-voltage problems found in the ignition leads and at the plugs. The two plug suppressors have metal ends for clipping to the plug head and the connector at the end of the HT lead, while the lead suppressor is in-line, designed to be inserted between two halves of the lead. You'll need a sharp knife . . .

then the engine earth strap. If the integrity of these is in any doubt whatsoever sling them out.

You might have realised that a metal bonnet acts as a kind of shield between engine and antenna. In some cases the effect of this can be enhanced by fitting a braided earth lead between the chassis of the car and the bonnet hinges at the point they fix to the bonnet itself, with beneficial effects on interference.

As a last resort you can follow the example of the Army. Very often, if not always, their radio trucks use an extreme form of suppression whereby all the HT leads are covered with a braided metal wire which screens them remarkably effectively. If your car is of the GRP (Glass Reinforced Plastic) type this final step may well be necessary, although before that you can try fitting a thin sheet of metal to the underside of the bonnet and earthing it with a metal strap—this will nearly always produce the desired results. In a like manner, owners of this type of car may well find that it is useful to mount the rig itself inside a metal box which is separately earthed. Alternatively they may well decide to sell the car and buy a real one.

With GRP cars in particular, although the theory applies to metal jobs as well, it is no bad idea to mount the antenna as far away from the engine as possible. The penalty of a more complex installation together with a longer feedline is outweighed by the decrease in potential interference.

Now, if you've done everything properly, you should find that your rig is working very nicely thank you, and all will be well. In order to make sure that it stays that way there are certain elementary things which you can do every so often—perhaps when the car is serviced—which will make sure that it all stays that way. The obvious things are to clean and check all the connections—especially earths, and including the normal vehicle earths—as well as the battery terminals. Also check all the wiring for breaks or wear. The antenna—especially the retractable sort—will benefit from being kept clean and free from rust and other natural forms of slime, and may also enjoy the occasional rub down with a mild spirit. The carbon-fibre HT leads, which are more common than the copper-core variety, tend to break down after a period of time—they are especially affected by heat. Make sure that they're not being baked by the exhaust manifold or decomposing because of contact with oil or petrol or similar. If they are then they must be replaced. Apart from affecting your rig such deterioration will also be affecting the performance of the vehicle engine.

Remember also that your rig does not like transmitting when the antenna is not connected or—and this is particularly relevant to retractable masts—when it is not the correct length. If it is retracted or bent, broken or otherwise deficient you will cook the output transistors in the rig. Expensive.

As a final courtesy to the world you might wish to examine the possibility of the alternative type of interference. This is the variety which emanates from your rig and affects others rather than them affecting you. Fellow breakers will doubtless tell you if your signal is poor or affecting neighbouring channels as well as the one you're transmitting on, in which case you must take the set to a radio engineer qualified to repair it. In the States such engineers must be licenced by the FCC, and it is an offence for anybody else to twiddle with the internal gubbins.

Also, before you rush off into breakerland transmitting like crazy, try parking the car next to, or as close as possible to, a TV set, preferably belonging to someone you know. Pressing the transmit button a few times on diverse channels will soon show up on the screen if your set is radiating TVI. Fitting a low-pass filter to the rig will eliminate this with no bother, and is a polite thing to do. Such interference on the TV actually indicates that the TV itself is lacking in some way, and although modern sets have filtering built-in it is also possible to buy a high-pass filter for connection to the TV if yours suffers a lot from spurious TVI.

Aside from bits which relate to improving reception on your rig in this way there are a whole load of add-on gadgets available, some of which are more effective than others and some of which are simply more expensive than others.

Power Mike

If your rig is of the sort which doesn't have adjustable mike gain you can quite easily buy a power mike in the local CB shop. This contains a small battery and has its own volume control which you can adjust (preferably with an on-air check) in the same way as with a built-in mike gain. It will have exactly the same results, perhaps better. A good power mike will almost always improve on the regular mike which came with the unit, especially as it will very likely be of the crystal type rather than the dynamic microphones which are normally supplied with a rig.

It will have to be connected to the rig in the usual way, via a connector, which will not be attached to the mike lead when you buy it. You will have to buy a connector and solder it on—as a rule instructions for accomplishing this on almost any make of rig will be included with the mike.

Beat Frequency Oscillator (BFO)

Remembering all the things we've discovered about radio waves and especially Sideband, you'll know that regular AM sets cannot receive signals from SSB. This is because the production of Sideband leaves the carrier out. A BFO puts it back again, enabling AM users to receive SSB signals. It does not convert an AM rig to full SSB operation.

Speech Compressor

This device plugs into the system between the microphone and the rig. You won't speak any differently into a speech compressor, or sound any different to people around you, but it will compress the electrical pattern the microphone has made from your voice into a tighter band, which in turn will give you better modulation on air. Your transmitted signal will sound a whole lot better. Like a power mike, it is possible for this device to overcook your voice, with the result that you will sound distorted. Some compressors are adjustable, and have a meter on the face to indicate level so that you can set them correctly to avoid over-modulation.

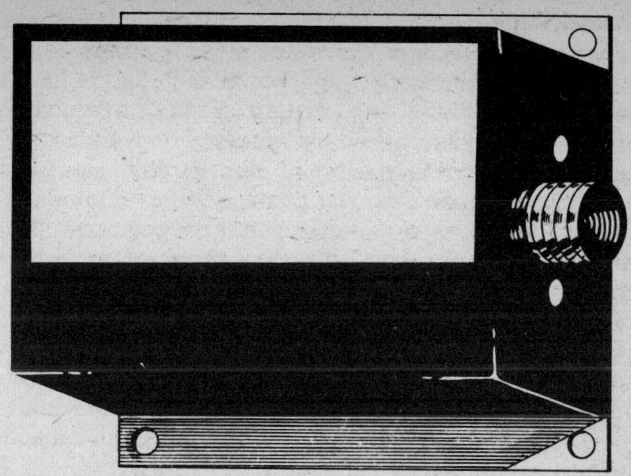

A close-up of an SWR meter. You can't see the dial scale because reflections are easier for artists to draw than numbers, but you can see that this one has been made with a backplate drilled for screws to facilitate permanent installation in the vehicle or at home.

SWR Meter

The use of this handy little tool should be no problem to anybody. Some meters are fairly substantial devices, with more than one dial scale. Some of them may even be of dual or treble role, including facilities for measuring SWR, power and signal strength. As a rule these are quite bulky objects, but frequently, and obviously more so with the smaller, single function sets, they are left in line constantly, so that SWR may be checked at will.

External Speakers

Most CB rigs have a loudspeaker built in to the unit. Frequently this will be muffled when the set is in place behind the dash, or pointed directly at the floor of the vehicle when it's below dash installation. An extra speaker, for which there is almost always a connection plug provided on the rig, may be mounted in a more practical location and can make the difference between understanding a weak signal and missing it completely.

PA Speaker

These have already been mentioned, and little further description of their application is needed. They are of plastic or light alloy construction in the shape of the traditional horn and are generally best mounted under the bonnet of the car, although exponents of the growing offroad trend may well find that mounting them outside on roof or nudge bars actually enhances the appearance of the vehicle rather than detracts from it.

Field Strength Meter

This measures radiated power in the area around the antenna, showing it either on a graduated scale of some sort or by buzzing at you. It can be used to show the pattern of radiation from your antenna simply by walking round the car, and will show you if and where your signal is being masked by the vehicle.

Dummy Load

The fact that attempting to transmit while the antenna is not connected will damage the rig has more than one drawback. For example anyone testing a rig after installation is going to be blotting out airspace which others may wish to use for talking on. To save annoying everybody else when you're testing late at night use a dummy load. This is nothing more than a 50–Ohm resistor which takes the place of the antenna, allowing you to transmit without transmitting and without the transistorised disaster which would normally accompany this act.

Tone Squelch

Although it's not as selective as the calling device now available through the miracle of the micro chip etc, tone squelch is a triggering device which responds to a musical tone and opens the squelch on your rig so that you may receive an incoming call from another rig similarly equipped. It gives you a certain amount of peace on the days when you don't want to listen to the CB, but ensures that you won't miss calls from your home base or your bank manager, or whoever else has got one. Specification is quite important with these, as some of the tones to which they respond are not far removed from sounds created in everyday speech patterns, and they may be almost useless.

Variable Frequency Oscillator
The use of the slider has already been discussed with reference to illegal use, and since we don't want to encourage that sort of thing we won't say another word about them. You rascals.

Frequency Counters
This device does exactly what its title implies; it reads your transmit frequency to several decimal places and displays it on a digital readout like a calculator. This is exceptionally useful for knowing if your set is on song, but only if you can be bothered to look up the exact frequency you should be on for any given channel. Remember there are 40 of them. Probably this is best suited to base station use, as motorists tend not to have very much time for this sort of thing. Not live ones, anyway.

Preamplifiers
This box of tricks goes in the feedline between antenna and rig, from which vantage point it will amplify all the signals coming into your receiver, thus making everything louder. The theory is wonderful but the practice is regrettably exact; it makes everything louder, indiscriminately, so that if you have a bit of ignition crackle on your rig in normal mode you will have a great deal of it with the preamp. They are thus less popular with mobile users than with base units, in which latter arrangement they can be very useful, particularly if they incorporate tuneable gain control, which is not a feature of all base units. Such an amplifier has its own power source

and is constantly in operation. It will therefore incorporate circuitry which prevents it operating for transmission.

Linear Amplifier
A thoroughly naughty device which boosts output to illegal levels. The less you know about these the better. Also available are bilateral amplifiers which boost both transmit and receive. They are also illegal. Like linear amps they are far more commonly available and used than they should be.

Birdcalls
These are also widely available, despite the fact that the FCC made them illegal in the States some while ago. They whistle and hoot into the airwaves. Their original justification (flimsy, to say the least) was that they attracted attention. They can drive you insane if you weren't already.

7

Base stations and antennas

Base stations are the next logical step up from a mobile rig for anyone who's having a good time on the air. Some of you may object to them being described thus, and the argument that CB provides the bulk of its benefits to the mobile user may well be justified, but in order to enjoy such benefits the mobile user is often dependent on base stations for local information and for response to emergencies.

In any case performance of a base station will almost always be better than that of a mobile unit. This is partly because the base set won't suffer from the electrical garbage which accompanies the successful operation of the petrol engine, nor will it be affected by limitations of size. And because the operator of a mobile set has to keep looking at the road in order to avoid the necessity for channel 9 broadcasting, the mobile set is much simpler than the base set, with fewer of the available refinements included.

The biggest single advantage which the base station enjoys is the superior nature of the antenna and its installation, and it is this which gives it nearly all of the extra range and clarity.

Purpose-built base units will operate on mains voltage and need only a three-pin plug to work correctly. The use of three-pin, to include the earth, is important here and should not be neglected.

Base units tend to be two or three times the price of mobile rigs and it may be that this factor will deter those people whose prime purpose for installing a rig at home is one of convenience rather than for the extra pleasure of indulging in a hobby. It is quite feasible, under such circumstances, to use a mobile rig at home as a base unit, although certain precautions have to be observed. First remember that it cannot be connected directly to mains voltage. You have two choices. One, and the most convenient, is to buy a transformer at almost any electrical shop, which will convert 240 volts AC into 12 volts DC. The transformer will provide its output through two wires, one positive and one negative. Make sure that these are connected to the rig in the correct way otherwise the resulting pyrotechnics will be expensive and dangerous.

Your other choice is simply to connect the rig directly to a car battery. Doing this will either mean that you have a spare battery or that you won't be driving your car any more. The major advantage of this is that you will still be on the air if there's a power cut. However you'll be the only one, so it won't do you much good. You will also have to put up with having a car battery lying about in your house. This smelly object (really) is unnecessarily heavy and also full up with corrosive juice which it will distribute all over your Persian rugs quite unaided by any human agency. It will be happiest if you disconnect it when the rig is not in use, which means that every time you switch on you will have to reconnect, paying especial attention to getting the wires the right way round. While nearly all of us can get such connections right once, having to make them three or four times a day will inevitably lead to the eventual cock-up extraordinaire. The battery will also need constant recharging, with further wiring complications, and constant monitoring in order to discover when such charging operations are necessary.

Doubtless you will draw your own conclusions from all this, and eventually arrive at some form of purchasing decision. In reality the thing which is going to make the most difference to the effectiveness of your base set will be the antenna, and we can happily spend hours arguing about the relative merits of different types. In fact a lot of people do, and you've probably copied them on channel before now. The fact is that the permanent installation offers far more scope for indulging in specialisation than the mobile antenna, which is essentially a general-purpose device giving the best all-round performance with consequent losses in particular areas.

The direct counterpart of the mobile antenna is the quarter-wave. This is the simplest form of base antenna and will do a basic job. Just how you improve on this depends more on what you wish to do with your base set than anything else. Because you can improve. The performance of an antenna is measurable in a roundabout sort of way and is called gain. The amount of gain which an antenna is capable of is expressed in decibels (db), but can be quite confusing. An antenna with 10db gain is not twice as good as an antenna with 5db gain. What happens is that if you run a 4 watt rig through an antenna with 3db gain it gives a result similar to running an 8 watt rig through an antenna with no gain at all. In order to calculate all this the basic quarter-wave antenna is considered to have no gain whatever, and the performance of

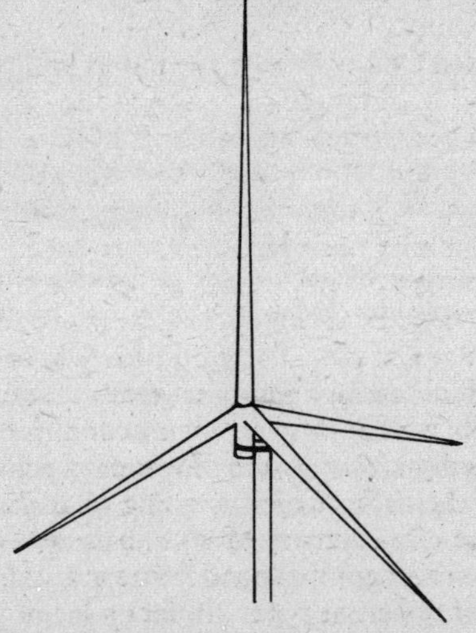

A quarter-wave vertical antenna of the kind which can easily be mounted on a pole. The radials at the bottom direct the signal downwards and give it some gain at ground level.

all other types is measured against this. A mast with 3db gain is then twice as effective as one with no gain.

When you buy an antenna for your base rig there may be many factors which you want to take into account, but most of all you should establish its gain. This figure will also be affected by the type of antenna you look at, since gain has to come from somewhere. A ground-plane antenna is so constructed as to inhibit your signal from radiating upwards and to use the spare power gained from this to send it further at ground level. This is clearly of no value if you wish to communicate with aircraft, balloonists or people who live in exceptionally tall buildings, but is the most common form of omnidirectional antenna.

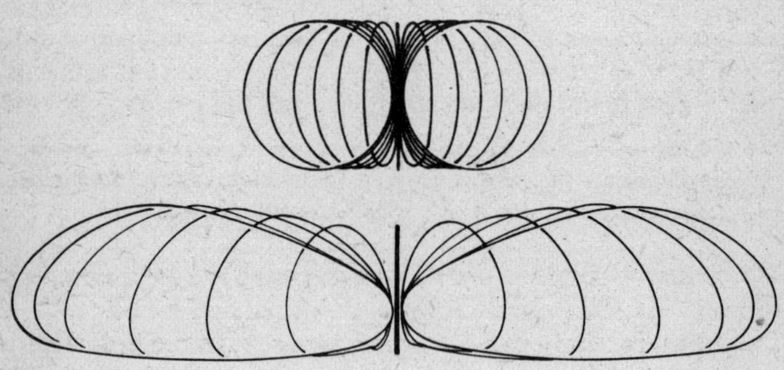

Gain. The difference in radiation pattern between the vertical antenna (top) and the ground-plane (bottom) is obvious, and the extra signal is gain.

The quarter-wave ground plane antenna, then, is the equivalent of a mobile whip with a bit extra and its main advantage is that it is light, strong and easy to set up and maintain. It will spread your signal equally well in all directions and is best suited to general-purpose uses. The next step is to go for a five-eighths-wave ground-plane which is the same only taller. The extra length of the driven element (the antenna itself) means that it functions better than a quarter-wave, and may have a 3 or 4db gain over the shorter mast. These basic types may vary in actual construction and appearance, par-

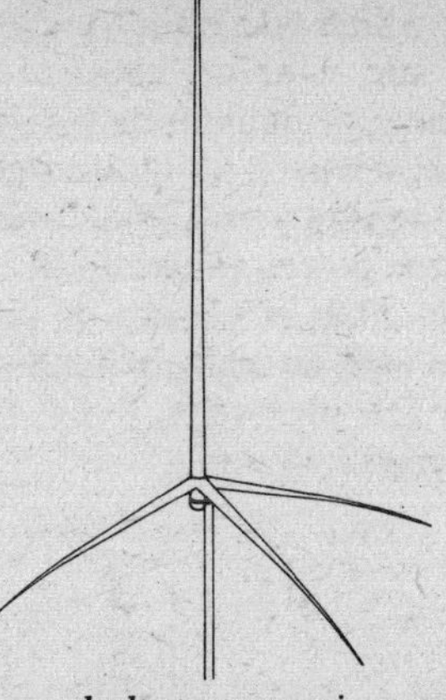

The five-eighth-wave ground-plane antenna is exactly the same in appearance as the quarter-wave. It's just twice the size, although not twice as effective. It is a definite improvement though.

ticularly between manufacturers, and the best comparison will be their respective gain measured against cost.

A directional antenna would be your next logical step from this, and it will have a much higher gain figure. This is because it uses a system of reflectors which will focus your signal in one direction at the expense of the others. If the directional antenna is pointed northwards your signal will travel a lot further, but only to the North. People to the South of you will have a lot of trouble picking you up at all.

The directional antenna requires a minimum of three elements to do this effectively. The middle one is the driven element, which is the one connected to the coax from the rig. Behind it is the slightly longer reflector, which is an electric mirror, if you like, rather like the dish used by radars and so on, which reflects the rearward-travelling signal forwards again. See? In front of the driven element is the director, which focuses the signal into a narrow band, giving it the strength to travel forwards. It is usually a little shorter than the driven element.

If you think of it like a car headlight the driven element is the bulb, the reflector is the silvered reflector and the director is the glass lens. Without the other two, the bulb would spread light in all directions, including up. Add the reflector and it points it all vaguely forwards. Add the lens and the light becomes a concentrated beam travelling in a clearly defined direction. It is easy to see that the directional antenna is much better if you want your signal to go places, but only if you are absolutely certain of where you want it to go. If you are using your base as a convenient means of communication

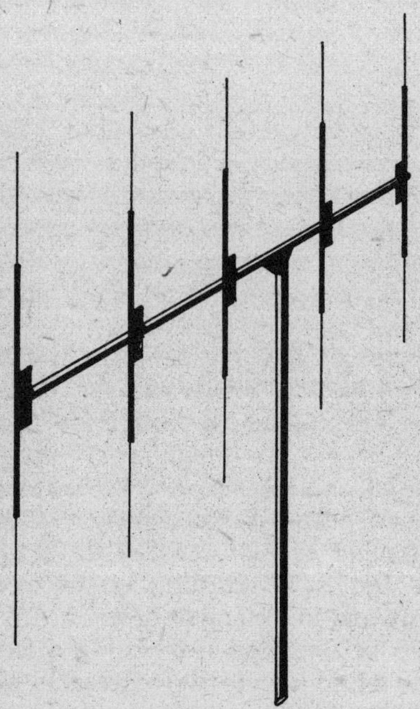

A five-element beam antenna. This, as you can see, is getting slightly cumbersome, but it has advantages which make up for the fact. It points towards the shortest element, although a good artist's use of perspective may well fool you.

between two fixed points this will be admirably suited to your needs. If you are not, however, you may find that further sophistication becomes necessary.

This sophistication will simply take the form of a rotator, which is nothing more than a little electric motor which will move your beam to point towards any place you choose. You can now speak to any number of fixed points at the flick of a switch. The size of the motor will depend entirely upon the size and weight of the antenna you choose, and this can vary enormously.

It's also worth mentioning that radio waves can be polarised either horizontally or vertically. Vertical waves travel with their crests moving up and down, horizontal waves from side to side. Obviously enough, mobile antennas are vertically polarised. This means that if you wish to use your base rig for general communication you would be better off with a vertically polarised antenna, since although vertical masts may receive horizontal signals and vice versa their ability to do so is not as great as if their polarisation were the same. This can be utilised for base to base communication, by using horizontal polarisation at each installation in order to cut down somewhat on unwanted reception of signals from mobiles in the area.

Given all this the most common type of directional antenna is the vertical beam, which works well for distance and short-range traffic. They are usually made out of relatively lightweight tubing, with consequent rewards in terms of installation simplicity and size of motor needed to rotate them. A three-element beam should have a gain of about 8db. The addition of further elements makes the arrangement more cumbersome, but has better gain. Four-element beams will give about 11db and five-element beams about 14db.

Some beams can be switched from being directional to being omnidirectional which clearly represents an advantage—the best of both worlds, in fact.

A stacked beam is one which has everything we've just mentioned, but more than once. Instead of being a beam on top of a mast it is a boom on top of a mast, with a beam at each

Horizontal and vertical polarisation explained. One picture (or in this case two) is supposed to be worth a thousand words. Say no more.

end of the boom. In fact you could go on indefinitely, or at least to the limits of your engineering and constructional ability. You will also need a very heavy-duty rotator to move these things around.

But the rewards are there, in terms of gain. If you use two three-element beams you will gain about 12db. Two four-element jobs will produce about 14db, while two five-elements will give you an astounding 17db—roughly equivalent to putting 200 watts into a basic quarter-wave mast.

Even more effective than this can be the gain from a quad antenna. These are square wire elements supported on a cruciform frame which have a much larger capture area than the straight elements. In fact a simple five-element quad can give as much gain—17db—as a twin stacked beam with five elements. Quads can be stacked also, and although a stacked quad doesn't double its gain the results are staggeringly high in terms of a total gain figure for the unit.

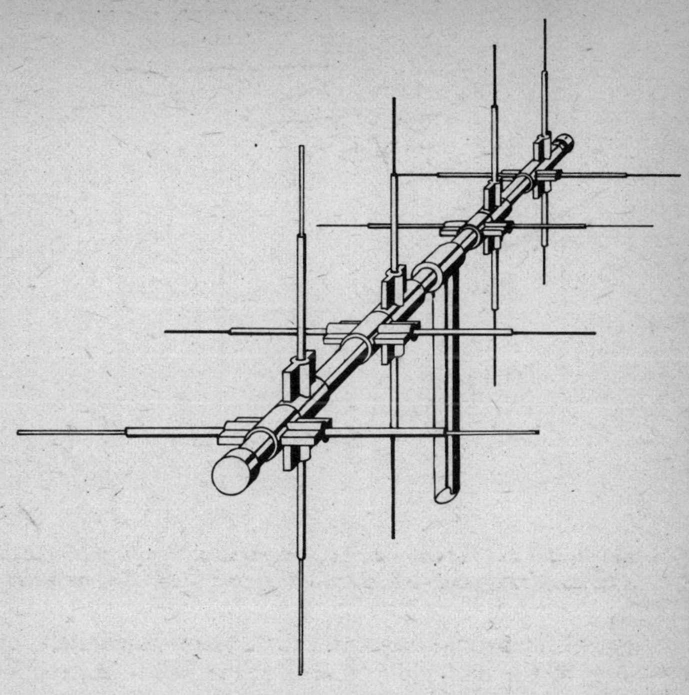

The big one. A complicated beam arrangement which may be switched from vertical to horizontal polarisation at the press of a button. Both vertical and horizontal modes are directional, and both go the same way.

All antennas will benefit from height when they are installed, and the higher up they go the better they will work. The total height you can achieve is limited by several factors, not least the local building regulations. Although the Eiffel Tower is a famous and much-admired historic landmark, half-size replicas in Surbiton may not be all that commendable. The Eiffel Tower also has the advantage of being remarkably sturdy in construction. Any tower you build for

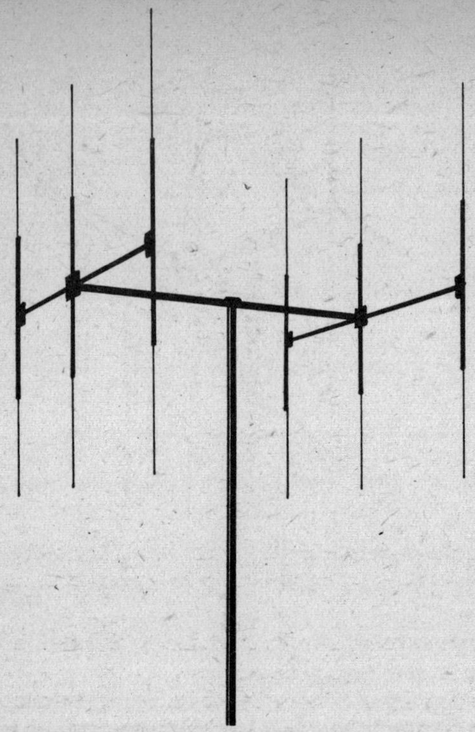

Stacked beams like this one can get really complicated and heavy. They also present quite a problem of perspective. Nevertheless they still point in the direction faced by the shortest elements.

your antenna must be erected with strength in proportion to the load it bears—stacked five-element beams will require a mast of some strength. In fact, after a point, a simple mast, however strong the central column and however deeply it is set in the ground, will be insufficient for the task, and you may well have to get into the tripod business in a big way, as well as performing excavations which the NCB could justifiably regard with some pride. Bear in mind also that when using directional antennas the rotator will be fitted at the top

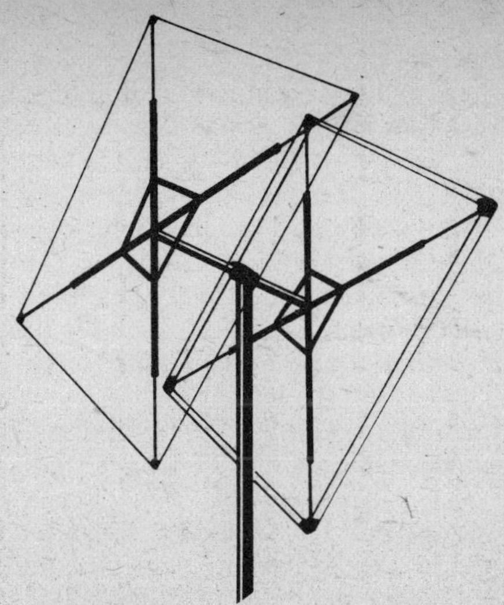

A quad. Immediate savings in weight are obvious from the use of wire. The greater catchment area of the quad is also easy to spot.

of the mast, turning the beams alone, and will add extra weight to the very top of the edifice.

Simpler fixing can be achieved by placing your antenna on a short mast fixed to your rooftop chimney or a straight pole run up the side of the house itself and bolted directly to the wall. This method, while it offers much in the way of simplicity, loses by way of detracting from performance. Proximity to objects like trees or chimney pots can mask transmissions in certain directions, without accounting for the interference between your CB antenna and the TV aerial which is very likely to be fixed to the same chimney.

Any free-standing mast or tower should be well bedded into the ground with a cement base to fix it. It should also be guyed with metal wires. Be sure that such wires are not fixed so close to the top of the tower that they interfere with the rotation of the beam. Also use insulators on the guy wires about every five feet or so until you are at least 25 feet away from the antenna, otherwise the wires themselves will degrade the signal quality.

When you install the rotator at the top of the tower you may wish to follow a fairly elementary precaution. Most of these devices are controlled by a switch which indicates which point of the compass the antenna is facing. This alignment does not happen by accident and you will have to do it yourself. When everything is in place you are well advised to check that the rotator can turn the beam in all directions and that nothing is fouling. Then rotate the beam until the switch indicates that the beam is pointing North. Then climb up the mast and adjust the beam until it is actually doing so. After that the beam will always be pointing where the switch says it is pointing, which you will find is exceptionally convenient.

Run the power lead for the rotator down the mast as far away from the feedline between rig and antenna as possible. Anchor it firmly and frequently, and ensure that it is not left loose in the open. Burying it is not recommended, and it may be necessary to support it on short poles or along a garden fence. When you lead it into the house and up to the switch don't cut it to exact length—you may want to move the rig about from time to time, so a few loops of spare cable will be handy.

When you run the actual feedline out to the antenna all rules which applied to mobile installation apply as well—doubly so. You must use a coax wire, otherwise you can give up straight away. No other kind is acceptable.

All electrical wiring offers a resistance to the current travelling along it, and coax is no exception. This resistance increases with length and will start to affect your rig badly once you're using a run of wire longer than 100 feet or so. Keep the feed to the antenna as short as you can, but if it goes over the 100 foot mark you would be advised to use a larger diameter coax. The thicker core of this will be of a lower resistance. It will also cost a middle-sized fortune.

Run the coax down the mast as far as possible—taking it off halfway down may be convenient for access to a bedroom window, but if it's too close to the beam you'll have a spot of bother since it will affect your signal adversely. As a general rule don't take the coax away from the mast until you're about 15 feet lower than the beam, preferably further.

If you have to join the coax make sure that you do so by means of connectors with joints soldered to the wire. If the connection has to be out in the open then it should be weatherproofed with plastic tape at least, preferably with a proprietary caulking gunge which will protect it but remain soft, so that the connection can be undone without resorting to brute force or explosive devices.

Such a joint may well be the ideal place for you to install an in-line lightning arrestor. While lightning is a pretty thing to watch at a distance it is also frighteningly destructive, having a capacity for damage at least as powerful as a train load of football fans. It is not the sort of thing which the average citizen would want in the living room, but unless you take preventive measures that is where it will end up. Sticking a lump of metal up on a pole is an invitation to lightning. Running a conductor (your coax) from the mast into your house is an act of folly comparable to running onto the terraces at West Ham wearing Tottenham colours.

A lighting arrestor will divert the lightning from the coax, utilising the fact that lightning will always follow the path of least resistance on its way to earth. A wire from the arrestor should be led in as straight a line as possible to a good earth—either a proper earth rod sunk at least six feet into the ground or to your cold water pipes. If you want the damn stuff in the house, that is. By the same token it is no bad idea to run the coax through at least one tight turn before you bring it indoors, and even with an arrestor it won't hurt, during electrical storms of any strength, to disconnect the lead at the rig and chuck it out of the window. It would also be a good idea to earth the tower or mast separately if it is not well bedded into the ground already. Any free-standing or rooftop mast should have a straight run of cable from its framework to an earth rod. Or else.

The last thing to do is to check the SWR of your antenna,

since the problem of dimension affects it in exactly the same way as with mobile antennas. Do this when the antenna is mounted on its mast, not before. Following the manufacturer's instructions for assembly will make a good SWR more likely, and you can check all the dimensions before you put the antenna up on its mast. The same rules for SWR apply with base sets as with mobiles and you must follow the same procedure. Be grateful if you get a figure significantly below 2:1, and if you can twiddle it to less than 1.5:1 (although it's not worth the bother)—give me back my wallet. Fagin.

8

The 10 code, the Q code and the CB insult code

The side of the CB world which seems to attract the most attention from the media and from the uninitiated is the jargon. The trucker-talk of America has metamorphosed into a strange private language made up of a compendium of words and phrases from several sources.

It has long been an accepted part of radio procedure that certain short cuts may be taken in airborne conversation both to cut down on the amount of time needed to complete a transmission and also to render it as unambiguous and simple to understand as possible.

To begin with there are several accepted codes to do this—Morse is one such, and although the need to spell out each individual letter of every word may seem time consuming and tedious it does at least eliminate the possibility of misunderstanding. Morse, however, was principally used because at the time there was no means of modulating a radio carrier in such a way that voice signals could be transmitted. As soon as this was possible it is arguable that Morse became outdated and obsolete, but it is still in widespread use today.

Using the spoken word can lead to unwanted people monitoring your transmissions and understanding perfectly things which perhaps you would rather remained uninterpreted. The equally pressing need for brevity has led to the establishment of various codes using letters, numbers or words.

The most relevant to CB was initially the Q code, which has the same meaning all over the world for a given series of three-letter codes all beginning with Q. Radio hams still use this extensively and it was also adopted for use on CB. A great deal of the original code proved to be reasonably superfluous to the needs of a short-range radio facility and so much of it goes unused. The Q code is now heard on CB relatively infrequently; in most cases it is used by Sideband operators more than anyone else and even then their conversations tend to consist of normal speech with the odd phrase from the code cropping up from time to time. The most commonly heard part of this code is QSL, with reference to little postcards send by radio users to confirm a radio contact with another user, particularly if long-distance communication is involved.

The thing which captured most imaginations was the 10 code in use by the various law enforcement agencies in the States. By establishing a number code preceded by the figure 10 they could communicate the fact that a major crime was in progress simply by giving a number over the air—in this case 10–35. Their code phrase for an acknowledgement was 10–4, and it was this curt ending to transmission growled into the mike by Broderick Crawford in the course of the TV series *Highway Patrol* which probably did most to establish it in the minds of the people who were later to become the CB afficionados.

Adoption of the 10 code on CB was more or less inevitable, although like the Q code it contained more than a passing number of items which were irrelevant to the everyday user not actually engaged upon the apprehension of desperate criminals. From this situation evolved the unofficial 10 code, used only on CB. Unfortunately this code, simply by virtue of being unofficial, seems never to have been recorded in exactly the same way more than once. For this reason there are several apparent duplications, where one piece of the code has two or more meanings, or where one message is represented by two separate ten-signals. In every case where this seems to have occurred we offer all accepted meanings or codes without preference; you'll have to work out for yourself which code to use in cases of conflict and which meaning to accept when a code has more than one. Jolly good luck indeed is all that remains to be said on that particular subject.

Fortunately for all of us the jargon is much simpler.

Unfortunately for all of us in England the jargon is still very largely based on that established in the States, principally by the truck drivers. Having said that it is worth mentioning that the CB jargon, like every kind of language, is essentially an evolutionary process, and as it is used more and more frequently in this country, by more and more people, it will gradually acquire a degree of Anglo-Saxon literalism which will probably make it a whole lot more acceptable to everyone. In the meantime, however, it is still very much the middle-American Country and Western argot of C W McCall, which needn't necessarily be all bad. It needn't necessarily be all good either, but we're stuck with it.

The official APSCO 10 code as used by law enforcement agencies:

10–0	Caution
10–1	Unable copy—change location
10–2	Signal good
10–3	Stop transmitting
10–4	Acknowledgement (OK)
10–5	Relay
10–6	Busy—stand by unless urgent
10–7	Out of service
10–8	In service
10–9	Say again
10–10	Fight in progress
10–11	Dog case
10–12	Stand by (Stop)
10–13	Weather—road report
10–14	Prowler report
10–15	Civil disturbance
10–16	Domestic problem
10–17	Meet complainant
10–18	Complete assignment quickly
10–19	Return to . . .
10–20	Location

10–21 Call . . . by telephone
10–22 Disregard
10–23 Arrived at scene
10–24 Assignment completed
10–25 Report in person (meet) . . .
10–26 Detaining subject, expedite
10–27 (Drivers) license information
10–28 Vehicle registration information
10–29 Check record for wanted
10–30 Illegal use of radio
10–31 Crime in progress
10–32 Man with gun
10–33 EMERGENCY
10–34 Riot
10–35 Major crime alert
10–36 Correct time
10–37 (Investigate) suspicious vehicle
10–38 Stopping suspicious vehicle
10–39 Urgent—use light, siren
10–40 Silent run—no light, siren
10–41 Beginning tour of duty
10–42 Ending tour of duty
10–43 Information
10–44 Request permission to leave patrol . . . for . . .
10–45 Animal carcass in . . . lane at . . .
10–46 Assist motorist
10–47 Emergency road repairs needed
10–48 Traffic standard needs repairs
10–49 Traffic light out at . . .
10–50 Accident
10–51 Wrecker needed
10–52 Ambulance needed
10–53 Road blocked at . . .
10–54 Livestock on highway
10–55 Intoxicated driver
10–56 Intoxicated pedestrian
10–57 Hit and run
10–58 Direct traffic

10–59 Convoy or escort
10–60 Squad in vicinity
10–61 Personnel in area
10–62 Reply to message
10–63 Prepare make written copy
10–64 Message for local delivery
10–65 Net message assignment
10–66 Message cancellation
10–67 Clear for net message
10–68 Dispatch information
10–69 Message received
10–70 Fire alarm
10–71 Advise nature of fire
10–72 Report progress on fire
10–73 Smoke report
10–74 Negative
10–75 In contact with
10–76 En route
10–77 ETA (Estimated Time of Arrival)
10–78 Need assistance
10–79 Notify coroner
10–80 Chase in progress
10–81 Breathalyser report
10–82 Reserve lodging
10–83 Supervise school crossing at . . .
10–84 If meeting . . . advise time
10–85 Delayed due to . . .
10–86 Office/operator on duty
10–87 Pickup/distribute cheques
10–88 Advise present telephone number of . . .
10–89 Bomb threat
10–90 Bank alarm at . . .
10–91 Pick up prisoner/subject
10–92 Improperly parked vehicle
10–93 Blockade
10–94 Drag racing
10–95 Prisoner/subject in custody
10–96 Mental subject

10–97 Check (test) signal
10–98 Prison/jail break
10–99 Records indicate wanted or stolen

The 10 code used on the CB band in America:

10–1 Receiving poorly
10–2 Receiving well
10–3 Stop transmitting
10–4 Acknowledged, message received
10–5 Relay message
10–6 Busy, stand by
10–7 Out of service
10–8 In service
10–9 Repeat message
10–10 Transmission completed
10–11 Talking too fast
10–12 Visitors present
10–13 Advise weather and/or road conditions
10–16 Make a pickup at . . .
10–17 Urgent business
10–18 Anything for me?
10–19 Nothing for you
10–20 Location
10–21 Call on landline
10–22 Report in person to . . .
10–23 Stand by
10–24 Completed last assignment
10–25 Can you contact . . .
10–26 Disregard last information
10–27 I am moving to channel . . .
10–28 Identify your station
10–29 Time is up for contact also I am leaving this location
10–30 Does not conform to FCC rules
10–31 Crime in progress
10–32 Radio Check
10–33 Emergency traffic at this station
10–34 Trouble at this station, help needed urgently
10–35 Confidential information

10–36	Correct time
10–37	Wrecker needed at . . .
10–38	Ambulance needed at . . .
10–39	Your message delivered
10–41	Please change to channel . . .
10–42	Traffic accident at . . .
10–43	Traffic holdup at . . .
10–44	I have a message for . . .
10–45	All units please report
10–46	Assist motorist
10–50	Break channel also traffic accident at . . .
10–51	Wrecker needed at . . .
10–52	Ambulance needed at . . .
10–53	Road blocked at . . .
10–55	Drunk driver
10–59	Convoy or escort
10–60	What is next message number?
10–62	Unable to copy, use landline
10–63	Network directed to . . . or prepare to make written copy
10–64	Network clear or network directed to . . .
10–65	Awaiting next message or network clear
10–66	Cancel message
10–67	All units comply
10–68	Say again
10–69	Message received
10–70	Fire at . . .
10–71	Proceed with transmission in sequence
10–73	Speed trap at . . .
10–74	Negative
10–75	You are causing interference
10–77	Negative contact or ETA
10–81	Reserve hotel room for . . .
10–82	Reserve hotel room for . . .
10–84	My telephone number is . . .
10–85	My address is . . .
10–88	Advise telephone number of . . .
10–89	Radio repairman needed at . . .

10–90 I have TVI
10–91 Talk closer to your microphone
10–92 Your transmitter is out of adjustment
10–93 Check my frequency on this channel
10–94 Please give me a long count
10–95 Please transmit a 5-second carrier
10–97 Check test signal
10–99 Mission completed
10–100 Personal reasons, rest-room stop
10–200 Police needed at . . .
73's Best wishes, regards
88's Love and kisses

Q code signals used by Ham operators internationally and also by Sidebanders to an extent:

QRA What is the name of your station?
QRB How far are you from my station?
QRC Will you tell me my exact frequency?
QRD Where are you going and where are you from?
QRE What is your estimated time of arrival at . . .?
QRF Are you returning to . . .?
QRH Does my frequency vary?
QRK What is the readability of my signals?
QRL Are you busy?
QRM Are you suffering from interference?
QRN Natural interference—static
QRO Increase transmitting power
QRP Decrease transmitting power
QRQ Transmit faster
QRS Transmit more slowly
QRT Stop transmitting
QRU Anything for me?
QRV Are you ready?
QRW Shall I inform . . . that you are calling him on channel . . .?
QRX When will you call me again?
QRY When is my turn?
QRZ Who is calling me?

QSA What is the strength of my signals?
QSB Are my signals fading?
QSL Acknowledge receipt
QSM Repeat last message
QSN Did you hear me on channel . . .?
QSO Communicate with . . .
QSP I will relay
QSW Do you wish to transmit on this channel?
QSX Listen to . . . on channel . . .
QSY Change to channel . . .
QSZ Send each word or sentence more than once
QTE What is my bearing from you?
QTH What is your position?
QTI What is your course?
QTJ What is your speed?
QTL What is your heading?
QTN Departure time?
QTR Correct time?
QTU Which hours is your station open?
QTV I will listen for you on channel . . .
QTX Listen for me until . . .
QUA Have you news of . . .?
QUD Have you received the emergency signal sent by . . .?
QUF Have you received the distress signal sent by . . .?
QUM Is the emergency traffic ended?
QUO Shall I search for . . .?
QUR Have survivors been picked up?
QUS Have you sighted survivors or wreckage?
QUT Is the position marked?

The CB insult code—the 13 code:
13–1 You're an idiot
13–2 I'm ignoring you
13–3 You're beautiful when you're angry
13–4 Sorry
13–5 And you!
13–6 So I made a mistake
13–7 If you can't copy it must be your fault, because I'm

running 2000 watts

13–20 Is your mike clinking or are your uppers loose again?

13–21 Are you being paid by the word?

13–22 Is that your voice or did you install a steam whistle?

13–23 If you had spoken for another 30 seconds you would have qualified for a Broadcast Station licence

13–24 You make more sense when you're pissed

13–25 The local operators have had a whip-round to buy your rig from you. Just stick to watching *Crossroads*

13–26 Next time you eat garlic speak further from the mike

13–40 Your signal sounds great. Now switch off your rig and call me on landline so I can find out what you want

13–41 Either my rig is out of adjustment or you're on channel 53

13–42 Either my speaker cone is ripped or you'd better try again when you're sober

13–43 That was a beautiful 10. Now try it with the mike connected

13–44 I love the way your new rig sounds. Now I know why that model was discontinued

13–45 Your transmitter must have a fault because there's smoke coming from my loudspeaker

13–46 That's a new antenna? I could get a better signal from a six-inch piece of damp string

13–47 What a fantastic signal. Give me a few minutes to bring the mobile unit to your driveway so I can copy your message

13–50 Can you slide that thing down 250KHz?

13–51 You've tried the upper Sideband, you've tried the lower side, you've even tried both. Now will you go QRT so we can use the central slot?

13–52 Best thing about SSB is that now you're only half as offensive as you were on AM

13–53 Attention AM stations on centre slot. Just because they won't talk to you on your own channels what makes you think we'll listen to you here?

The internationally recognised phonetic alphabet is in common use by radio enthusiasts—professional, amateur or CB—all the time, and is used frequently for spelling complicated words or in describing places, people or things by their initials:

ALPHA
BRAVO
CHARLIE
DELTA
ECHO
FOXTROT
GOLF
HOTEL
INDIA
JULIET
KILO
LIMA
MIKE
NOVEMBER
OSCAR
PAPA
QUEBEC
ROMEO
SIERRA
TANGO
UNIFORM
VICTOR
WHISKY
X-RAY
YANKEE
ZULU

9

Trucker-talk

ACE a CB operator with a high opinion of himself
ADIOS goodbye
ADVERTISING police car with flashing lights and siren
AFFIRMATIVE yes
A LITTLE HELP extra power—a linear amplifier
ALLIGATOR CB operator who doesn't reply
ANCHOR MAN base station operator also anchored modulator
APPLIANCE OPERATOR CB user who knows nothing about his rig

BACK replying, as in COME BACK, BRING BACK etc
BACK BENDER Sideband user
BACK DOOR last vehicle in a convoy
BACK DOOR CLOSED rear of convoy covered for police
BACKGROUND interference
BACK OFF stop transmitting or slow down
BACK OFF THE HAMMER slow down
BACKSIDE return trip
BACKSTROKE return trip
BAD SCENE crowded channel
BAGGING police catching speeding motorists
BALLET DANCER antenna blowing in the wind
BAREFOOT using a rig legally, without extra power
BASEMENT channel 1
BASE TWENTY operator's home
BEAN STORE restaurant
BEAR police

BEAR BAIT speeding vehicle without CB
BEAR CAGE police station
BEAR CAVE police station
BEAR IN THE AIR police helicopter or aeroplane
BEAR REPORT information on the location of the police
BEATING THE BUSHES scout vehicle ahead of convoy going fast enough to attract police attention and draw them from concealment
BEAVER small furry animal with sharp teeth
BEAVER BREAKER female on channel
BEAVER HUNT searching for beaver. Once bitten, twice shy, eh?
BETTER HALF spouse
BETWEEN THE SHEETS sleeping
BIG C North Circular Road
BIG CIRCLE North Circular Road
BIG SWITCH the on/off switch on a CB rig
BIG 10–4 agree one hundred per cent
BLACKTOP major road
BLEEDING transmission affecting channels either side of the actual one in use
BLESSED EVENT a new CB rig
BLOOD WAGON ambulance
BLOWING SMOKE loud clear signal
BLOW THE DOORS OFF overtake
BOBTAIL artic tractor unit without a trailer
BODACIOUS loud signal
BOOTLEGGER operator without a licence
BOOTS a linear amplifier
BOULEVARD main road
BRA BUSTER lady with huge norks
BREAKER a CB user
BREAKING THE NEEDLE coming in loud and clear
BROWN BOTTLES beer
BRUSH YOUR TEETH AND COMB YOUR HAIR warning, radar ahead
BUBBLE GUM MACHINE vehicle with flashing lights, usually a police car

BUBBLE TROUBLE a tyre problem
BUCKET MOUTH CB user who talks a lot on air, usually using bad language
BUMPER LANE overtaking lane on motorway
BURNER a linear amplifier
BURNING MY EARS loud clear signal
BUST getting caught by the law for speeding, illegal transmission etc
BUTTON PUSHER CB user who presses the PTT bar but never speaks
BUZBY GPO monitors

CABOVER type of lorry which is flat-fronted
CAMERA police radar
CARRIER radio wave with no voice transmission
CHECK THE SEAT COVERS look at the passengers
CHECKING MY EYELIDS FOR PINHOLES sleeping
CHICK woman or girl
CLEANER CHANNEL channel with less interference
COME AGAIN repeat last transmission
COME ON bring it back etc
COME BACK bring it back and so on
COMIC BOOK trucker's log book
COMING OUT THE WINDOWS good loud reception
CONVOY line of vehicles in CB contact travelling together
COPY receive a transmission
COPYING THE MAIL listening to a CB conversation without speaking
COUNTRY CADILLAC farm tractor
CUT THE COAX turn off the rig
CUT SOME Zs get some sleep
CRADLE BABY new CB user who's afraid to ask someone to stand by
CREDIT CARD boyfriend, male passenger with female breaker

DEAD KEY person pressing the PTT bar for some time without speaking thus blanking out the channel

DEAD PEDAL slow moving vehicle
DESPAIR BOX place where spare CB components are kept
DIARRHOEA as in verbal diarrhoea—breaker who won't shut up
DIESEL DIGIT channel 19, the truckers' channel
DO IT TO ME answer, come back, come on, etc
DON'T FEED THE BEARS don't get caught speeding
DOUGHNUT tyre, also more recently, a roundabout
DOWN AND GONE turning off the CB
DOWN AND ON THE SIDE transmission over but listening
DO YOU COPY? do you understand?
DARKTIME night
DRAGGING WAGON breakdown truck
DRESS FOR SALE prostitute
DROP THE HAMMER accelerate

EARS CB radio or antenna for same
EARWIG listen to a conversation
EASY CHAIR vehicles in the middle of a convoy
EIGHTS goodbye, but with love and kisses. Use this farewell with caution
EVEL KNIEVEL motorcycle policeman
EVERYBODY IS WALKING THE DOG channels are all busy
EYEBALL meet someone or look at something
EYETIES the Italians, who all have their own CB rigs and use them

FEED THE BEARS get a speeding ticket
FINGERS a channel hopper
FIVE as in, give me a five, transmit numbers 1 to 5 for a radio check
FIVE BY FIVE perfect signal. Also, KICKING FIVES or READ YOU IN FIVES
FLAG WAVER motorway maintenance worker
FLAT SIDE sleep
FLIP-FLOP return trip

FOLDING CAMERA Visual average speed computer and recorder (VASCAR)
FOOT WARMER linear amplifier
FOR SURE agreed
FOUR as in THAT'S A FOUR means yes
FOUR TEN emphatically yes
FOUR-WHEELER car
FOXHUNT meanies looking for CB
FRONT DOOR leading vehicle in a convoy
FUNNY BOOKS porno magazines
FUR-LINED SEATCOVER female passenger

GET HORIZONTAL go to sleep
GLITCH an indefinable defect
GOING DOWN switching off the CB rig
GO JUICE petrol or diesel fuel
GOLDILOCKS a mobile businesswoman
GOOD BUDDY a fellow CB user
GOOD LADY the same but a lady
GOOD NUMBERS best wishes
GOON SQUAD group who hog the calling channel
GOT A COPY? can you hear?
GOT MY EYEBALLS PEELED looking
GO TO 100 head for a rest-room stop
GREEN STAMPS money
GROUND-CLOUDS fog

HAMMER accelerator
HANDLE nickname by which CB user is known on air
HANG A LEFT/RIGHT make a left/right turn
HARVEY WALLBANGER reckless or bad driver
HIGH GEAR using a linear amplifier
HIT THE HAY got to sleep
HOLDING ON TO YOUR MUDFLAPS driving close behind you
HOLE IN THE WALL tunnel, area of bad reception
HOME PORT home address
HOME TWENTY home address

HOTEL OSCAR the Home Office
HOT WIRE linear amplifier

IDIOT BOX television set
INDIAN as in Tennessee Valley Indian = neighbour suffering TVI from you
INVITATION speeding ticket or summons

JAILBAIT very young girl
JAM SANDWICH marked police traffic car
JAW JACKING talking
JUICE petrol or diesel fuel

KEEP THE SHINY SIDE UP AND THE GREASY SIDE DOWN have a safe trip
KEEP YOUR NOSE BETWEEN THE DITCHES AND SMOKEY OUT OF YOUR BRITCHES have a safe trip
KEYBOARD controls on a CB set
KEYING THE MIKE pressing the transmit button without speaking
KICKER linear amplifier
KODAK police radar set
KOJAK WITH A KODAK policeman operating radar

LADY BEAR policewoman (not only Angie Dickinson)
LANDLINE telephone
LAY AN EYEBALL ON meet in person, go to see
LET THE CHANNEL ROLL let others use the channel
LETTUCE money
LINEAR AMPLIFIER illegal device to boost transmit power higher than the permitted maximum
LOAD OF VW RADIATORS empty truck or empty container of any sort

MAIL overheard conversation
MAKING THE TRIP getting your signal out
MAMA BEAR lady police officer
MAYDAY internationally recognised distress signal

MEANIES anti-CB authorities
MEAT WAGON ambulance
MIKE microphone
MOBILE EYEBALL check out a car or truck while moving
MOBILE PARKING LOT car transporter
MODULATE talk
MONSTER LANE nearside lane on a motorway
MOTION LOTION petrol, diesel fuel
M-20 meeting place

NEGATIVE CONTACT no answer to a call
NEGATORY no
NINE repeat last transmission as in GIVE ME A NINE

ON CHANNEL on the air
ONE-EYED MONSTER television set
ONE-TIME short contact
ON THE PEG at the legal speed limit
ON THE SIDE monitoring the channel, listening watch
OPEN on the air
OTHER HALF wife, husband, girlfriend etc
OUT transmission completed, no further contact expected, as in FINAL
OVER transmission ended, your turn to speak
OVERMODULATING muffled or distorted signal
OVER YOUR SHOULDER behind you

PANIC IN THE STREETS area being monitored by the Post Office
PAVEMENT PRINCESS prostitute
PEANUT BUTTER IN YOUR EARS not listening to CB
PERMANENTLY 10–7 dead
PICTURE BOX police radar
PIECE OF PAPER speeding ticket
PLAIN WRAPPER unmarked police car
PLAY DEAD stand by or refuse to answer
POLAROID radar
PORTABLE FARMYARD cattle truck

PORTRAIT PAINTER policeman operating radar
POSITIVE yes, affirmative
PREGNANT ROLLER SKATE VW Beetle
PRESS THE SHEETS go to sleep
PRESSURE COOKER sports car
PULL THE BIG SWITCH go off the air
PUT AN EYEBALL ON look at, meet
PUT THE HAMMER DOWN accelerate
PUT THE PEDAL TO THE METAL accelerate, travel flat out

Q CODE internationally recognised set of code signals
QSL CARD postcard sent to confirm CB contact, usually over a long distance
QUICK TRIP AROUND THE HORN scan all the channels for activity

RADIO CHECK report on quality and strength of transmission
RATCHET JAW CB operator who talks too much
READING THE MAIL eavesdropping
REEFER refrigerated trailer
RELOCATION CONSULTANT removal van
RIG truck or CB set
RINGING YOUR BELL someone calling you
RIOT SQUAD neighbours suffering TVI and complaining
ROAD TAR coffee
ROCK crystal used to tune CB rig
ROGER yes, affirmative, also ROGER DODGE, ROGER D
ROLLER SKATE production car
RUBBERBANDER novice CB operator
RUNNING BAREFOOT using legal power for CB

SAIL BOAT FUEL wind or an empty petrol tank
SALT SHAKER salt spreading vehicle
SAVAGES breakers hogging the channels
SEAT COVER passenger, often female

SEEING EYE DOG radar detector
SETS OF DIALS CB rig
SEVENTY-THREES best wishes, regards
SHAKE THE TREES AND RAKE THE LEAVES vehicle at rear and front of a convoy watching for police in wait ahead or catching up from behind
SHAKING THE WINDOWS loud and clear
SHOTGUN passenger
SHOES linear amplifier
SIDEWINDER SSB user
SITTING UNDER THE LEAVES concealed police car
SIX-LANE CAR PARK motorway
SKATING RINK slippery road
SKY HOOK antenna
SLAMMER jail
SLIDER illegal device used to transmit between authorised channels
SMOKEY the police
SMOKEY ON RUBBER police vehicle
SMOKEY REPORT information on location of police
SMOKEY WITH A CAMERA police using radar
SMOKEY WITH EARS police with CB
SNAFU cock-up
SOCKS linear amplifier
SPAGHETTI Italian CB transmissions
SPARKS electrician
SPLIT YOUR SIDES transmit on Sideband
SQUAWK BOX CB rig
STACK THEM EIGHTS best regards
STEPPED ON YOU interrupted your transmission
STEREO coming in loud and clear
STINGER antenna, especially top-loaded variety
STROLLER operator on foot with a hand-held rig
SUCKER CB rig in for repair
SUICIDE JOCKEY truck driver carrying explosives or petrochemicals
SUPERSLAB motorway
SWEET THING lady on channel

SWINDLE SHEET truck driver's log book

TAKE IN THE SLACK accelerate
TAKE IT UP/DOWN move to a specified channel
TAKING PICTURES police using radar
TEAR JERKER breaker with a sob story
TEN FOUR yes
TEN FOUR HUNDRED drop dead
TEN POUNDER excellent radio
THERMOS BOTTLE tanker truck
THIN weak signal
THIRTY THREE emergency signal
THIRTY TWELVE definitely agreed—ten four three times over
THROWING transmitting
THROWING A CARRIER transmitting a carrier wave as in KEYING THE MIKE
THROWING NINE POUNDS coming in loud and clear
TIGER IN THE TANK linear amplifier
TIGHTEN UP THE RUBBER BAND accelerate
TIN CAN CB rig
TOILET MOUTH breaker using obscene language
TWENTY location
TWO-WAY RADAR police using radar from a moving vehicle
TWO WHEELER motorbike

WALKED ALL OVER blotted out by another transmission
WALKING TALL coming in loud and clear
WALKING THE DOG talking skip
WALLPAPER QSL cards
WALL TO WALL loud and clear, as in WALL TO WALL AND TREE-TOP TALL
WALL TO WALL BEARS police everywhere
WALL TO WALL SPAGHETTI overwhelming skip from Italy
WALLY an idiot

WEARING BOOTS/SOCKS/SHOES using a linear amplifier
WHAT'S YOUR TWENTY? where are you?
WHEELS a vehicle, also set of wheels
WINDOW WASHER rain, as in GOD'S WASHING THE WINDOWS
WOBBLER vehicle without CB
WRAPPER the colour of a vehicle—in a blue wrapper, red wrapper etc
WRINKLE a hitch or blank spot during transmission

X-RAY MACHINE police radar
XYL wife, from Ex Young Lady

YL girlfriend, from Young Lady
YOU GOT IT the channel's yours, go ahead, agreed
YO-YO vehicle varying its speed

ZOO police station

10

TECHSPEAK—a glossary of CB terminology

Although we've tried to keep everything as simple as possible throughout it has been impossible to completely eliminate the need for technical terminology. They should all have been explained along the way, but this list ought to clear up any that have been missed. Once again it's not an attempt to explain the theory of relativity in words of one syllable, but it should cover the basics.

AC Alternating Current, as used in mains electricity
AGC Automatic Gain Control. Built in to a CB rig it maintains constant volume on both strong and weak stations
AM Amplitude Modulation—a way of putting a voice signal onto a radio wave
AMPERE Unit of electrical current
ANL Automatic Noise Limiter. Accessory on a CB rig which reduces interference from vehicle ignition and other sources
ANTENNA The aerial to which a CB rig is connected
APA Aerial Pre-Amplifier boosts incoming signal
ATU Aerial Tuning Unit—a fine tuning device

BAND A group of frequencies or channels
BASE STATION Transceiver at a fixed location
BEAM Directional type of CB antenna

CARRIER The radio wave which carries voice signals
CHANNEL Specific frequency within a radio band
CLARIFIER Control on Sideband sets which helps make the incoming signal intelligible

COAX Shielded cable that is used in antenna feedlines
CRYSTAL Device used to tune the CB exactly to a chosen frequency

DC Direct Current as used in motor vehicles
DECIBEL Unit of sound measurement used to rate antennas for performance
DELTA TUNE Control on CB rig for fine tuning stations slightly off channel
DUAL CONVERSION Receiver circuit to cut out interference from stations using adjacent channels
DUMMY LOAD Device which turns radio waves into heat and allows running a transmitter without an antenna for off-air testing
DX Long-distance exchange of radio signals

EARTH Connection to the ground itself, or a common wiring harness through a chassis in a car or radio set
ELEVEN-METRE BAND Another way of describing the 27MHz CB band. 11 metres is the wavelength of 27MHz CB

FET Field Effect Transistor—high quality transistor used in the more expensive rigs
FIELD STRENGTH METER Instrument which measures the strength of a radio signal, useful for tuning and antenna positioning
FREQUENCY SYNTHESIS Circuit which reduces the number of crystals necessary for all-channel reception
FM Frequency Modulation—a way of putting voice onto a radio carrier which is more refined than AM

GAIN An increase in power
GMRS General Mobile Radio Service
GROUND PLANE Nondirectional or omnidirectional antenna for base station use. It has one vertical driven element and several radial rods around the base

HANDSET Combined microphone and earphone like a telephone

IC Integrated Circuit. A one-piece chip holding thousands of transistors. Several entire sections of a transceiver

IMPEDANCE The opposition or resistance to a flow of electric current. 27MHz CB sets have an impedance of 52ohms throughout for a perfect match

JACKPLUG A single pin plug for audio connections like the loudspeaker or PA horn

LINEAR AMPLIFIER Simple but powerful amplifier to boost output of a CB rig to an illegal level

LOADED ANTENNA Antenna shorter than wavelength but with a loading coil somewhere in its construction

LOADING COIL Wire wound coil used to increase the electrical length of an antenna

MHz MegaHertz. Measure of how many times a radio wave vibrates in a second. Mega means million and Hertz is cycles per second. A 27MHz signal therefore vibrates 27 million times per second

MICROVOLT One millionth of a volt

MIKE An abbreviation for microphone, also sometimes seen as "mic"

MODULATING Vibrating a radio carrier with the voice

NOISE BLANKER Device similar to an Automatic Noise Limiter but more sophisticated

OHM Measurement of electrical resistance

OMNIDIRECTIONAL The sending of a signal in all directions. Also described as Nondirectional

OUTPUT POWER The power generated by a transceiver or other radio/audio device, measured in watts

PEP Peak Envelope Power is used to measure the power output of Sideband sets

PLL Phase Lock Loop. Like a synthesiser this circuit in the rig allows the transmission of signals on many frequencies

without the need to have a crystal for each channel
PTT Push-to-talk bar on a radio mike allows transmission when pressed and reception when released

QUARTER-WAVE ANTENNA The basic CB antenna, which can be used for base or mobile installations

RF Radio Frequency. Electrical field which vibrates fast enough to create radio signals
RX The receive stage of a transceiver
RIG Slang word for a CB transceiver

SELECTIVITY The ability of a receiver to ignore interference and transmissions not exactly on the frequency it is tuned to
SENSITIVITY The ability of a CB set to receive weak, distant transmissions and amplify them to the point where they become audible
SIGNAL STRENGTH Strength of an incoming signal measured on a sale of 1–9 on an S-meter
SKIP Phenomenon whereby short-wave radio signals are reflected by the ionosphere
S-METER Scaled dial on a CB rig to show strength of received transmissions
SQUELCH Control on a CB set used to eliminate noise from interference when no signal is being received
SSB Single Sideband. An efficient means of sending a radio signal which makes better use of the available channel allocations
SWR Standing Wave Ratio. A way of comparing and measuring forward and reflected power in an antenna system

TX The transmit stage of a radio transceiver
TRAFFIC Radio messages
TRANSCEIVER A radio set capable of transmitting and receiving radio signals
TRANSISTOR Tiny solid device which performs basic electrical functions

TVI Television Interference

VOLT Measurement of electric current or pressure
VOX Voice operated transmitter, which is activated by speech in the microphone rather than a PTT bar

WATT Measurement of electric power
WHIP An antenna

11

CB people—clubs, associations and REACT

Like many other things we all enjoy, CB is habit-forming. Sooner or later the dabblers end up hooked, the casual users dependent, the occasional listeners-in unable to exist without their radio. Like many other large groups of people with a common interest CB users tend to form themselves into clubs. To begin with they swap ideas, knowledge, experiences and equipment, plus various other odds and ends; it's probably not a coincidence that most clubs of this nature tend to hold their meetings in pubs—they're social events rather than boring think-ins.

The number of such clubs increases in direct proportion to the growth of interest in CB, and so does their membership; some clubs have sufficient members to be able to raise substantial sums of money for various charities and good causes—something which they nearly all seem to be doing at present.

There's probably no better way to get involved with CB than through a club like this; they can help you with basics like installation and operation of your rig, help to familiarise you with the CB scene in general and also show you a smart time into the bargain. At a local level you can't go wrong. Some of these clubs sprang originally from the campaign to get CB legalised in the UK and as such are organised on a regional and national level and offer perhaps a slightly higher degree of participation. If participation is what you're looking for then you would do well to consider the possibility of becoming involved with REACT. Formed in the USA in the sixties the REACT (Radio Emergency Associated Citizen's

Teams) ideal was to provide 24-hour monitoring of the emergency channel 9 by volunteer operators on a nationwide basis. This ambition has come as close to realisation as it is possible to be and today the American REACT teams enjoy a close and cooperative relationship with the traditional official emergency services. Their monitors, while volunteers, are well trained and efficient and have been responsible for bringing a great deal of help to people who needed it and indeed have saved life on more than a few occasions. REACT can always do with more help and the rewards in terms of personal satisfaction are likely to be immense. The REACT setup for the UK was organised very early on in the development of the infant (and pirate) CB facility and waited until the government Green Paper before emerging properly from the shadows. In view of their splendid record of cooperation with the authorities in the States REACT has been anxious to avoid close association with the pirates over here in order not to jeopardise the formation of similar close links with British emergency services and so far this policy seems to be working well. With luck we should be able to look forward to the same degree of assistance and protection from REACT monitors as CB users in the States.

In view of the nomadic and occasionally transitory nature of some of the smaller CB clubs information concerning their whereabouts and meeting places is likely to date very quickly—quite apart from the fact that any such list is almost certain to be incomplete. If the chairmanship of some clubs has passed on since this list was finalised the outgoing incumbent should still be able to put you in touch. Failing that, at least you know the club exists—details of its present location can probably be found quite easily from one of the magazines dealing with CB or from one of the nationally organised clubs which maintains a register of clubs—someone like the CBA would be good for this.

Other than that, and especially if you're already on the air, give a call on 19 almost everywhere except London, which uses 14 mostly and occasionally 20, or on Sideband try 16LSB which seems to be nationally accepted as the calling channel.

Anglia Breakers Club
c/o Great White Horse Hotel
Tavern Street
Ipswich
Suffolk

Bracknell Breakers
Every Sunday at
The Bridge House
Working Road
Bracknell
Berks

Breaker One Four Club
c/o OK Corral
Napier Barracks
BFPO 20
West Germany

Breakers Town CBC
Every Tuesday at
The Stanley Club
Stanley Road
Carshalton
Surrey

Bricket Breakers Club
c/o Watford Component
Centre
7 Langley Road
Watford
Herts

Bristol CBC
1a St Peters Rise
Headley Park
Bristol BSI3 7LU

Bristol Breakers
120 Beaufort Road
St George
Bristol 5

Bruggen Bandits CBC
on channel 14
West Germany

Bury CBC
c/o Ripley House Hotel
Northgate Avenue
Bury St Edmunds
Suffolk

Cardiff & District Breakers
412 Aberdored Road
Gabalfa
Cardiff

CBA Centgral
5 Carronvale Avenue
Larbert
Stirlingshire

CBA Fife
32 Balmaise
Leven
Fife

CBA Reading
PO Box 123
Reading

CBA Sussex
15 Buckingham Mews
Shoreham By Sea
Sussex

CBCB Club
103 Southwood Road
Downside
Dunstable
Beds

CBGB
CB House
Crosby
Liverpool

CB-NE
PO Box 61
Sunderland

CB Radio Action Group
55 Dartmouth Road
Forest Hill
London SE23

Cheesy Breakers Club
116 St Christopher's Drive
Caerphilly
Glamorgan

Cheltenham Breakers Assn
60/6 Pitville Crescent
Cheltenham
and
The Crown and Cushion
Bath Road
Cheltenham

City Circle Breakers Club
Bedford Green
Horseferry
Leeds

Clog Town Breakers Club
33 Pendle Court
Astley Bridge
Bolton BL1 6PY

Clyde Coast Breakers
c/o Island Hotel
New Street
Stevenston
Ayrshire

Copy Cats Club
The Manager
Martholme Grange
Altham
Accrington
Lancashire

Crewe Breakers Club
1 Main Road
Wynbunbury
Crewe
Cheshire

Derwent Valley Breakers
10 Prospect Terrace
New Kyo
Stanley
County Durham
and Wednesdays at
The Black Bull, Lanchester

Don Valley Breakers
15 Roseberry Avenue
Hatfield
Doncaster

Dragon Breakers Assn
96 Leafield Road
Hunts Cross
Liverpool 25

East Antrim CBRC
PO Box 4
Antrim

Edinburgh CBRC
22 Ross Gardens
Edinburgh EH9 3BR

Elite Breakers
The Windmill
Lambeth High Street
London SE1

Elite Breakers
The Father Thames
Albert Embankment
London SE12

Essex Citizens' Band Club
24 Byrony Close
Witham
Essex CMB 2XF

Farnborough Area Breakers (FAB)
Every Tuesday at
The Oasis Club
Alexander Road
Farnborough
Hants

Fluxton, Urnstan & Davy
White Good Buddies
Associates
P.O. Box 2
164 Corn Exchange Buildings
Manchester 4

GBA
Coronation Service Station
Middleton Road
Heywood
Lancs

Glasgow CBC
147 Trossachs Road
Glasgow G72

Gramplan Breakers Club
59 Jasmine Terrace
Aberdeen
Scotland

Harrow and Wembley CB Group
7 Sandringham Crescent
Harrow
Middlesex

Hazard County Breakers Club
22 Radcliffe Avenue
Chaddesden
Derby

Independent Breakers Assn
113 Biscot Road
Luton

Lagan Valley CB Club
Poste Restante
GPO
Lisburn
N Ireland

Leicestershire CB'ers
c/o Modern Motoring
68 Narborough Road
Leicester LE3 0BR

Lennox Breakers' Club
Loch Lomond
Dunbartonshire

Leslie Breakers
Mondays at
The Leslie Arms
Cherry Orchard Road
Croydon

Lorn Breakers Club
4 Lismore Crescent
Orban
Argyll

Medway Breakers
56 Playstool Road
Newington
Sittingbourne
Kent

Meon Valley Breakers
4 Lawrence Road
Fareham
Hants

Merseyside 27 Club
34 Micklefield Road
Liverpool 15

Mexico City Breakers
The Old Masons Arms
High Street
Mexborough
Yorks

Mid-Kent CBC
c/o Ten Four Telecom
22 The Broadway
Maidstone
Kent

Midlands CBRC
Unit 2
72 Oval Road
Erdington
Birmingham

Midlands CB Radio Club
85 Allens Lane
Pelsall
Walsall
West Midlands

Milktown Breakers
Every Wednesday at
Rawthorpe Working Mens
Club
Rawthorpe
Huddersfield

NACB
Every Thursday at
The Commerce International
Nuthall Road
Nottingham

National Committee For
The Legislation of 27MHz
CB Radio
47b Stoneygate Road
Narborough
Leicester

National Independent Pirate
Band
Heirman Straat 37
Merksem 2060
Belgium

Northampton Breakers Club
Wednesdays & Sundays at
The Needle
Northampton

North Notts Breakers
38 Williams Street
Langold
Worksop
Notts

Open Channel CBC
17 Coronation Street
Preston

Open Channel Citizens
Band Club
17 Coronation Street
Blackburn

Open Channel Club
87 Sedgemoor Road
Coventry CV3 4EA

Pennine One Nine Club
29 Legrams Avenue
Lidget Green
West Yorkshire

Preston CBC
29 Russell Avenue
Preston
Lancs

REACT Area 2—South
West
28 The Coots
Stockwood
Bristol
Avon BS14 8LH

REACT Area 3—South
East
56 Parsonage Road
Cranleigh
Surrey GU6 7AJ

REACT Area 4—Midlands
130 Chester Road South
Kidderminster
Worcs DY10 1XE

REACT Area 5—North &
I. of Man
7 Brendjean Road
Morecambe
Lancs LA4 5SE

REACT Area 6—Scotland
10 Manse Road
Stonehouse
Lanarkshire ML9 3QP

REACT Area 7—Wales
10 Hendre Close
Capel Hendre
Ammanford
Dyfed

REACT Area 8—N Ireland
Representation pending

REACT Area 9—Greater
London Area
130 Drysdale Avenue
North Chingford
London E4 7PE

Redditch Area CB Club
88 Herondfield Close
Churchill
Redditch
Worcs

Richmond & District
Breakers
Friday nights at
Black Horse
Richmond

St Helens CB Club
33 Broadway
Grange Park
St Helens
Merseyside WA10 3RS

Sandwell Area CB Club
4 Baldwin Close
Twidale
Warley
West Midlands

Seven Towers CBC
15 Carnduff Drive
Ballymena
Co Antrim

South Birmingham Citizens'
Band Club
14 Delrene Road
Shirley
Solihull
West Midlands

South Wales Big 10–4 Club
12 Elgin Street
Manselton
Swansea

Stag Town Breakers Club
Every Thursday at
Courtlands Social Club
Thorpe Road
Bellamy Road Estate
Mansfield
Nottingham

Steel City CBC
282 Eccleshall Road
Sheffield S11 8PE

Swindon CBC
c/o 23 Affleck Close
Toothill
Swindon SN5 8DF

10–4 Club
85 Essex Road
Walthamstow
London E17

21 Breakers Club
c/o 8 Trerice Drive
Tretherras
Newquay
Cornwall

UBA (NE)
53 Mayfair Avenue
Lancaster

UKCBC
32 Downbank Avenue
Barnehurst
Kent DA7 6RP

National Committee for the
Legislation of CB
32 Downbank Avenue
Barnehurst
Kent DA7 6RP

United Breakers Association
50 Gaskel Street
London SW4

United Campaign For The
Legalisation of CB Radio
10 Lochnell Road
Dunbeg
Connel
Argyll RA37 1QJ

Untouchables
299 Manchester Road
Kearsley
Bolton
Lancs

Wessex Open Channel Club
48 Holsom Road
Stockwood
Bristol BS14 8LX

West Glamorgan Breakers
Assn
25 Plass Newydd
Baglan Moors
Port Talbot
West Glamorgan SA11 7DF

West London Breakers
meet at the Steam Packet
by Kew Bridge

West London Breakers
Tuesday at
White Hart
Southall

Weymouth CBC
Flat 1
39 St Thomas Street
Weymouth
Dorset

Wyre Forest Breakers
19 Chawson Pleck
Chawson Estate
Droitwich

Young Breakers Association
22 Romerby Crescent
Bolton
Lancs

12

CB-UK—map and town handles

The practice of describing towns—not all of them major—by some means other than the name by which they are more commonly known is yet another American habit which seems to have been imported along with the CB rigs themselves. However, as yet relatively few such town handles exist at all. Of those that do some are very obvious and refer to a landmark or characteristic of the place which is so well-known and long established that it was probably being taught in Geography or History lessons when I was at school, which only goes to show that I'm older than you think. I'm probably older than I think as well.

In the very earliest days of the boom in CB piracy in this country some of the first exponents of the art, and later on the bulk of them, were the youngsters who were also members of the Street Clubs. This provided CB with a close-knit fraternity who were already used to being looked on as slightly strange by the community in general and also, by virtue of the eyecatching cars they drove, quite used to being regarded with deep suspicion by the forces of law and order. CB probably wouldn't have thrived as swiftly as it did had it not been for this unofficial nation-wide federation with their own grapevine and a previously established and well-organised habit of gathering together in large groups on almost every summer weekend. It's a communications network of such widespread capability and having such a low profile that it would make the average KGB spymaster drool in anticipation.

Apart from this semi-clandestine organisation many of

the clubs had already adopted titles for themselves based on the area they came from—Leicester Rod and Custom Club is a practical, if unimaginative example of such an appellation. Among all this straightforward nomenclature lurked several clubs who had named their town, for the purposes of naming their club, in a style which was later to reveal itself as being remarkably suitable for CB operators. This we can only assume to have been a lucky accident, but is useful nonetheless.

Even so it's reasonably early days for all this and the list of town handles is by no means complete. Nor is it free from controversy. In the end the only successful handle for a town is one which the breakers are happy to use, and if they don't like it they won't have anything to do with it, no matter how erudite or amusing it may be. This is amplified in a few cases where towns seem to have got themselves handles which are neither erudite, amusing, nor even informative, but are downright uncomplimentary, to say the very least.

More than a few towns seem to have inherited handles based simply on the first letters of their name, which is all very well until you find that seven other places are known by the same title. Then you get confusion, or in extreme cases, fisticuffs, until the matter is resolved. Which could be never.

Bearing all this in mind, and having eliminated, where possible, cases of doubt, when preparing this map we were left with a remarkably short list. But we felt that it was still vaid enough for inclusion, so here it is.

There are also a few counties which seem to have earned themselves handles as well, so we've slung those in for good measure (they are denoted on the map by the large size numerals). If you don't agree with the handle of one or several towns, or if you can make up another six of your own—don't worry about it. It's an evolutionary process which no individual is likely to influence to any large degree, so the best thing is to sit back and let it happen. It's going to anyway.

	Town name	*Handle*
(1)	Aberdeen	Granite City
(2)	Barking	Eastside
(3)	Belfast	Soul City
(4)	Birmingham	Dead City
(5)	Bolton	Chip Butty Town
(6)	Bridgewater	Smelly Town
(7)	Bury	Black Pudding City

(8) Coleraine	Bay City
(9) Crewe	Rail Town
(10) Croydon	Big C
(11) Edinburgh	Castle City
(12) Glasgow	Second City
(13) Gloucester	Big G
(14) Great Yarmouth	Golden Mile
(15) Hereford	Cider City
(16) Heywood	Quarry Town
(17) Horley	Night City
(18) Lincoln	Cathedral City
(19) Liverpool	Murky River City
(20) London	Smokey Town, also Smoke City
(21) Luton	Dreamy Town
(22) Manchester	Rainy City
(23) Morecambe	Bay Town
(24) Norwich	Fine City also Canary Town
(25) Oldham	Milltown
(26) Portrush	Surf City
(27) Portsmouth	Booze City
(28) Preston	Cotton Town
(29) Salisbury	Spire City
(30) Sheffield	Steel City
(31) Skelmersdale	New Town
(32) Sleaford	Customville
(33) Taunton	Cider Town
(34) Torquay	Sunny Bay City
(35) Winsford	Salt City
(36) Buckinghamshire	Royal County
(37) Cornwall	Surf County
(38) Devon	Cream County
(39) Humberside	Bridge County
(40) Kent	Hop County
(41) Northampton	Shoe County
(42) Somerset	Apple County

GENERAL FICTION

		CYRIL ABRAHAM	
Δ	042616184X	**The Onedin Line: The High Seas**	80p
Δ	0426172671	**The Onedin Line: The Trade Winds**	80p
Δ	0352304006	**The Onedin Line: The White Ships**	95p
Δ	042697114X	**The Shipmasters**	80p
		BRUCE STEWART	
Δ	0352305738	**The Onedin Line: The Turning Tide**	£1.25
		TESSA BARCLAY	
	0352304251	**A Sower Went Forth**	£1.50
		JUDY BLUME	
	0352302712	**Forever**	75p*
		ANDRÉ BRINK	
	035230703X	**A Dry White Season**	£1.95
	0352305916	**Rumours of Rain**	£1.95
	0352306904	**An Instant in the Wind**	£1.95
		MICHAEL J. BIRD	
Δ	0352302747	**The Aphrodite Inheritance**	85p
		ADRIAN BROOKS	
	0352307773	**The Glass Arcade**	£1.50*
		MAGDA CHEVAK	
	0352303514	**Splendour in the Dust**	£1.50
		JACKIE COLLINS	
Δ	0352395621	**The Stud**	£1.25
	0352300701	**Lovehead**	£1.25
	0352398663	**The World is Full of Divorced Women**	£1.25
Δ	0352398752	**The World is Full of Married Men**	75p

* Not for sale in Canada. • Reissues.
Δ Film & T.V. tie-ins.

GENERAL FICTION

	ISBN	Author / Title	Price
		CATHERINE COOKSON	
	0426163796	**The Garment**	95p
	0426163524	**Hannah Massey**	95p
	0426163605	**Slinky Jane**	95p
		HENRY DENKER	
	0352396067	**The Physicians**	95p*
	0352300523	**A Place for the Mighty**	75p*
	0352303522	**The Experiment**	£1.50*
		MEL ELLIS	
Δ	0352306386	**Wild Horse Hank**	95p*
		NORMAN GARBO	
	0352305665	**Cabal**	£1.50*
	0352304995	**The Artist**	£1.50*
		ROBERT GROSSBACH	
Δ	0352307528	**Chapter Two**	£1.25*
		ELIZABETH FORSYTHE HAILEY	
	0352304359	**A Woman of Independent Means**	£1.25*
		SUSAN HILL	
Δ	0352307242	**Breaking Glass**	95p
		BURT HIRSCHFELD	
	0352398582	**'Father Pig'**	95p*
	0352395176	**Secrets**	95p*
	0352398604	**Behold Zion**	95p*
		JOYCE JOHNSON	
	0352306718	**Bad Connections**	£1.25*
		SARAH KERNOCHAN	
	0352304200	**Dry Hustle**	£1.25*
		ELLIE LING	
	0352306416	**This Year's Girl**	£1.25
	0352304154	**The First Splash**	75p
		PAT McGRATH	
	0352304383	**People in the Crowd**	95p

* Not for sale in Canada. • Reissues.
Δ Film & T.V. tie-ins.

GENERAL FICTION

ISBN	Author / Title	Price
	MARIA ISABEL RODRIGUEZ	
0352305622	**Maestro of Alhora**	**£1.25**
0352303913	**The Olive Groves of Alhora**	**75p**
	JUDITH ROSSNER	
0352396946	**To The Precipice**	**85p***
0352302089	**Nine Months in the Life of an Old Maid**	**75p***
0352301465	**Any Minute I Can Split**	**95p***
	LAWRENCE SANDERS	
0352302135	**The Pleasures of Helen**	**95p***
	JEREMY SCOTT	
0352307048	**Hunted**	**£1.50**
	ALAN SILLITOE	
035230698X	**The Storyteller**	**£1.50**
0352300949	**Men, Women and Children**	**75p**
0352300981	**Saturday Night And Sunday Morning**	**£1.25**
0352395141	**The Widower's Son**	**£1.50**
0352397144	**The Flame of Life**	**£1.50**
0352398809	**The Ragman's Daughter**	**£1.25**
035230202X	**A Start in Life**	**£1.75**
0352301821	**Raw Material**	**95p**
0352302518	**Key to the Door**	**£1.50**
0352303263	**The Death of William Posters**	**£1.50**
0352303379	**A Tree on Fire**	**£1.35**
0352305185	**Guzman go Home**	**£1.25**

* Not for sale in Canada. • Reissues.
Δ Film & T.V. tie-ins.

GENERAL FICTION

	ISBN	Author / Title	Price
	0352306378	KELLY STEARN **Consequences**	£1.25
Δ	0352302968	KATHLEEN TYNAN **Agatha**	75p*
	035230183X	DOROTHY UHNAK **Policewoman**	75p*
Δ	035230684X	GERALD WALKER **Cruising**	95p*
Δ	0352304421	DAVID WEIR **The Water Margin**	£1.50
Δ	0523411774	BRONTE WOODWARD & ALLAN CARR **Can't Stop the Music**	£1.25
	035230779X	CHELSEA QUINN YARBO **Dead & Buried**	£1.50

GENERAL NON-FICTION

	ISBN	Author / Title	Price
	0352307781	FRED SCHRUERS **Blondie**	£1.25*
	0352307099	PEGGY ANDERSON **Nurse**	£1.50*
	0352307498	DENYS VAL BAKER **The Spirit of Cornwall**	£1.50
	0352301392	LINDA BLANDFORD **Oil Sheiks**	£1.50
	0352307501	LT.-COL. JOHN BLASHFORD-SNELL WITH MICHAEL CABLE **In The Wake of Drake**	£1.25
	0352396121	ANTHONY CAVE BROWN **Bodyguard of Lies** **(Large Format)**	£2.50*
	0352307005	ROBIN COLLYNS **Prehistoric Germ Warfare**	£1.25
	0352306432	RODNEY DALE & IAN WILLIAMSON **Myth of the Micro**	£1.50
	035230345X	RODNEY DALE AND JOAN GRAY **Edwardian Inventions**	£2.95
Δ	0352301368	JOHN DEAN **Blind Ambition**	£1.50*
	0352304413	MARGOT FONTEYN **A Dancer's World (illus)**	£1.95
	0426178653	RICHARD GARDNER **The Tarot Speaks**	50p
	0352303247	H. R. HALDEMAN **The Ends of Power**	£1.25*

* Not for sale in Canada. • Reissues
Δ Film & T.V. tie-ins